Atom Resolved Surface Reactions
Nanocatalysis

RSC Nanoscience & Nanotechnology

Series Editors

Professor Paul O'Brien, *University of Manchester, UK*
Professor Sir Harry Kroto FRS, *University of Sussex, UK*
Professor Harold Craighead, *Cornell University, USA*

This series will cover the wide ranging areas of Nanoscience and Nanotechnology. In particular, the series will provide a comprehensive source of information on research associated with nanostructured materials and miniaturised lab on a chip technologies.

Topics covered will include the characterisation, performance and properties of materials and technologies associated with miniaturised lab on a chip systems. The books will also focus on potential applications and future developments of the materials and devices discussed.

Ideal as an accessible reference and guide to investigations at the interface of chemistry with subjects such as materials science, engineering, biology, physics and electronics for professionals and researchers in academia and industry.

Titles in the Series:

Nanotubes and Nanowires

C.N.R. Rao, FRS and A. Govindaraj, *Jawaharlal Nehru Centre for Advanced Scientific Research, Bangalore, India*

Nanocharacterisation

Edited by A.I. Kirkland and J.L. Hutchison, *Department of Materials, Oxford University, Oxford, UK*

Atom Resolved Surface Reactions: Nanocatalysis

P.R. Davies and M.W. Roberts, *School of Chemistry, Cardiff University, Cardiff, UK*

Visit our website at www.rsc.org/nanoscience

For further information please contact:
Sales and Customer Care, Royal Society of Chemistry, Thomas Graham House, Science Park, Milton Road, Cambridge CB4 0WF, UK
Telephone +44 (0)1223 432360, Fax +44 (0)1223 426017, Email: sales@rsc.org

Atom Resolved Surface Reactions
Nanocatalysis

P.R. Davies and M.W. Roberts
School of Chemistry, Cardiff University, Cardiff, UK

RSCPublishing

ISBN: 978-0-85404-269-2

A catalogue record for this book is available from the British Library

© The Royal Society of Chemistry 2008

All rights reserved

Apart from fair dealing for the purposes of research for non-commercial purposes or for private study, criticism or review, as permitted under the Copyright, Designs and Patents Act 1988 and the Copyright and Related Rights Regulations 2003, this publication may not be reproduced, stored or transmitted, in any form or by any means, without the prior permission in writing of The Royal Society of Chemistry, or in the case of reproduction in accordance with the terms of licences issued by the Copyright Licensing Agency in the UK, or in accordance with the terms of the licences issued by the appropriate Reproduction Rights Organization outside the UK. Enquiries concerning reproduction outside the terms stated here should be sent to The Royal Society of Chemistry at the address printed on this page.

Published by The Royal Society of Chemistry,
Thomas Graham House, Science Park, Milton Road,
Cambridge CB4 0WF, UK

Registered Charity Number 207890

For further information see our web site at www.rsc.org

"It is not the critic who counts, not the man who points how the strong man stumbled or where the doer of deeds could have done better. The credit belongs to the man that is actually in the arena whose face is marred by dust, sweat and blood, who strives valiantly, who errs and comes short again and again, who knows the great enthusiasm, the great devotions and spends himself in a worthy cause, who at the best knows in the end the triumph of high achievement and who at the worst, if he fails, at least fails while daring, so that his place shall never be with these cold timid sorts, who know neither victory nor defeat."

Theodore Roosevelt

Preface

"We build too many walls and not enough bridges"

Isaac Newton

Since the earliest days of scanning tunnelling microscopy (STM) and the award of the Nobel Prize to Binnig and Rohrer in 1986, it was evident that a powerful approach had become available for the study of structural aspects of solid surfaces. The technique provided a surface probe with atomic resolution with initial emphasis being given to semiconductors, Binnig and Rohrer being at IBM. It was clear, however, that it would be applicable for the study of surfaces over a wide range of scientific disciplines with nanotechnology, over the last decade, developing into a "stand-alone" discipline with far-reaching implications in many areas of science and technology and with significant financial investment in both academic and industrial laboratories. Matthew Nordon of Lux Research, a nanotechnology consultancy in New York, suggested in *The Economist*, 1 January 2005, that in 2004 the American government spent $1.6 billion on nanotechnology, more than twice as much as it did on the Human Genome Project. Also that IBM had in 2005 more lawyers than engineers working in the field of nanotechnology relating particularly to patent rights. However, at that time there had been no law suits as no real money had been made from the nanoparticles – mainly carbon nanotubes.

Much of the early STM studies of metals was focused on surface restructuring, particularly adsorbate-induced changes. Interpretation of adsorbate images was initially by simple inspection but it became clear that information from other experimental techniques would be required for particle identification. *In situ* chemical information, to complement structural information, was generally lacking until the mid-1990s and this was a serious disadvantage if we were to probe the details of molecular events associated with a surface-catalysed reaction. There was also a tendency in surface science to study either single adsorbate systems or to carry out experiments where two adsorbates were introduced to the solid surface sequentially rather than simultaneously. The latter coadsorption approach could be argued to simulate more closely a "real catalytic reaction" than the former; it was an approach we had adopted at

Cardiff using surface-sensitive spectroscopies and which provided a different insight into surface reactivity, particularly of the role of transient and precursor states in the dynamics of surface-catalysed reactions.

Although this book is research oriented, we have attempted to relate the information and concepts gleaned from STM to the more established and accepted views from the classical macroscopic (kinetic, spectroscopic) approach. How do well-established models stand up to scrutiny at the atom resolved level and do they need to be modified? We have, therefore, included a chapter where classical experimental methods provided data which could profit from examination by STM.

In taking this approach, someone new to the field of surface chemistry and catalysis can hopefully obtain a perspective on how more recent atom resolved information confirms or questions long-standing tenets. There is, therefore, a historical flavour to the book, with the first chapter dealing briefly with "how did we get to where we are now?". This inevitably means that the views expressed reflect personal perspectives but are very much influenced by the outstanding contributions from those who have pioneered the development of STM in surface chemistry and catalysis, of which groups at the Fritz-Haber Institut in Berlin and the universities at Aahrus, Berkeley and Stanford have been at the forefront.

Thirty years ago, one of us (M.W.R.) set out with Clive McKee to write the book *Chemistry of the Metal–Gas Interface* (Oxford University Press, 1979; Russian translation, Moscow, 1981). This was prompted by the then rapid developments in surface-sensitive spectroscopies – infrared, photoelectron and Auger – and structural information from low-energy electron diffraction. The present book represents a step-change in the quest to understand surface phenomena through the wealth of information now becoming available from STM. The atom resolved evidence brings into focus the limitations of long-held views in surface chemistry and, where appropriate, we have given some hints as to what questions should be addressed when formulating reaction mechanisms and the challenge of providing meaningful kinetic expressions for surface reactions.

J.W. Mellor, in his book *Modern Inorganic Chemistry*, published by Longman Green and Co., made the following interesting observation:

> "*The word catalysis itself explains nothing. To think otherwise would lay us open to Mephistopholes' gibe: A pompous word will stand you instead for that which will not go into the head.*"

He then goes on to state that there is no difficulty in covering an obscure idea so that the word appears to explain the idea. This, written in 1920, still has a ring of authenticity nearly 100 years later, with Mephitopholes' gibe having a much wider relevance than being confined just to the word *catalysis*.

There is good evidence that STM has already and will continue to have a significant and far-reaching impact on our understanding at the molecular level of the dynamics and structural aspects of adsorption processes and their role in surface-catalysed reactions. There has, however, been only limited evidence on STM's impact in industrial or applied catalysis, the evidence being more obvious in materials science with atomic force microscopy (AFM) being used to greater advantage. There is indeed the view that the development of catalysts in the chemical industry has and will continue to rely very much on empirical skills in catalyst preparation, where nanoscale particles have been central to the control of both activity and selectivity. A knowledge of the development of principles and concepts which emerge from fundamental studies will, however, undoubtedly continue to influence thinking in industrial laboratories.

We hope that the book will appeal not only to those who wish to become familiar with the contribution that STM has made to the understanding of the field of surface chemistry and heterogeneous catalysis, but also to those who are new to catalysis, a fascinating and important area of chemistry and where so much has still to be achieved. Chapters have, where appropriate, suggestions for further reading where topics have been considered in more depth by others in both original papers and monographs. In addition to the references included with each chapter, dates are occasionally mentioned in the text to enable the reader to glean the time scale on which the science, concepts and experimental data first became available, then established and possibly modified.

About the Authors

Wyn Roberts, a student of the Amman Valley Grammar School, studied chemistry at University College Swansea where, after graduation, he pursued postgraduate studies investigating the role of sulfur as a catalyst in the formation of nickel carbonyl under the supervision of Keble Sykes. After being awarded his PhD, he was first appointed to a United Kingdom Atomic Energy Research Fellowship at Imperial College of Science and Technology, London, and then as a Senior Scientific Officer at the National Chemical Laboratory, Teddington. His first academic post was a lectureship at the Queen's University, Belfast, before in 1966 being appointed to the Foundation Chair of Physical Chemistry at the University of Bradford, where he also had periods as Head of Department and Dean of Physical Science.

In 1979 he moved to University College, Cardiff, where he was Head of Department (1987–1997), a Deputy Principal (1990–1992) and is currently a Research Professor. He was invited to be World Bank Visiting Professor in China in 1985, a Visiting Professor at Berkeley and is an Honorary Fellow of the University of Wales, Swansea. He was the first Chairman of the Surface Reactivity and Catalysis Group (SURCAT) of the RSC.

His research interests are in the application of surface-sensitive experimental methods in surface chemistry and catalysis and he has supervised over 80 PhD students, his co-author being one of them. He has received three National Awards, the Tilden Lectureship and Medal of the RSC, the Royal Society of Chemistry Award in Surface Chemistry and the John Yarwood Prize and Medal of the British Vacuum Society. He has also held appointments with the

University Grants Committee, the Science Research Council and the Ministry of Defence.

Phil Davies, a student of Blythe Bridge High School, Stoke-on-Trent, studied chemistry and mathematics in Southampton University. An undergraduate project on modelling of adsorption at fractal surfaces led to an interest in surface phenomena and, after graduating with double honours in 1986, he moved to Cardiff to study reactions at surfaces with surface-sensitive spectroscopy. After being awarded a PhD, he was appointed to a lectureship in the Department of Chemistry at Cardiff. His main research interests are studying reaction mechanisms at surfaces primarily through the use of surface sensitive spectroscopy but he also spent a short period of time with Professor Rutger van Santen in Eindhoven, studying adsorption and reaction using *ab initio* calculations on clusters. Since 1997, his interests have centred largely on the influence of local atomic structure on reaction mechanisms studied with scanning probe microscopies.

Between them, Davies and Roberts have published over 400 scientific papers and a number of books.

Acknowledgements

This book could not have been contemplated without the contributions that our graduate students have made to the understanding of molecular processes at metal surfaces, with Martin Quinn and Brian Wells providing early evidence from work function and photoemission studies of oxygen-induced surface reconstruction, Clive McKee and Richard Joyner of surface dynamics and structure, Julian Ross of "real" catalysis, Albert Carley and Paul Chalker in photoelectron spectroscopy and Chak-Tang (Peter) Au of surface transients in oxidation catalysis. These were studies that provided the science and impetus which led to the EPRSC funding an STM–XPS project in which Albert Carley, Giri Kulkarni, K.R. Harikumar and Rhys Jones were prominent. One of us (M.W.R.) has also enjoyed discussions over a wide range of surface chemistry with Ron Mason for some 30 years, exemplified by his prompting of a recent paper describing a novel approach to synthesising silica overlayers by nano-casting (Chapter 11).

We are also grateful for the permission we received from a number of groups to make reference (with figures) to their data, including Gerhard Ertl, Jost Wintterlin and Hajo Freund a the Fritz-Haber Institut, Bob Madix at Stanford, Gabor Somorjai and Miquel Salmeron at Berkely, the group of Besenbacker at Aahrus and Tøpsoe, John Thomas at Cambridge and Richard Palmer in Birmingham.

We also acknowledge the Institute of Physics' permission to make reference to quotations which appeared in *The Harvest of a Quiet Eye*, by Alan C. Mackey, Ed. M. Ebison, Institute of Physics, 1977.

Finally, we are indebted to Terrie Dumelow for her patience in preparing the manuscript and our families for their long-suffering support.

Contents

Abbreviations   xvi

Some Relevant Units   xvii

Chapter 1   **Some Milestones in the Development of Surface Chemistry and Catalysis**

1.1	Introduction	1
1.2	1926: Catalysis Theory and Practice; Rideal and Taylor	2
1.3	1932: Adsorption of Gases by Solids; Faraday Discussion, Oxford	2
1.4	1940: Seventeenth Faraday Lecture; Langmuir	2
1.5	1950: Heterogeneous Catalysis; Faraday Discussion, Liverpool	3
1.6	1954: Properties of Surfaces	4
1.7	1957: Advances in Catalysis; International Congress on Catalysis, Philadelphia	5
1.8	1963: Conference on Clean Surfaces with Supplement: Surface Phenomena in Semiconductors, New York	6
1.9	1966: Faraday Discussion Meeting, Liverpool	6
1.10	1967: The Emergence of Photoelectron Spectroscopy	6
1.11	1968: Berkeley Meeting: Structure and Chemistry of Solid Surfaces	7
1.12	1972: A Discussion on the Physics and Chemistry of Surfaces, London	7
1.13	1987: Faraday Symposium, Bath	8
1.14	Summary	8
	References	11
	Further Reading	12

Chapter 2 Experimental Methods in Surface Science Relevant to STM

2.1	Introduction	13
2.2	Kinetic Methods	13
2.3	Vibrational Spectroscopy	14
2.4	Work Function	15
2.5	Structural Studies	16
2.6	Photoelectron Spectroscopy	18
2.7	The Dynamics of Adsorption	21
2.8	Summary	26
	References	27
	Further Reading	30

Chapter 3 Scanning Tunnelling Microscopy: Theory and Experiment

3.1	The Development of Ultramicroscopy	31
3.2	The Theory of STM	35
3.3	The Interpretation of STM Images	37
3.4	Scanning Tunnelling Spectroscopy	38
3.5	The STM Experiment	40
3.6	The Scanner	42
	3.6.1 Sample Approach	43
	3.6.2 Adaptations of the Scanner for Specific Experiments	43
3.7	Making STM Tips	44
	3.7.1 Tip Materials	46
	References	48

Chapter 4 Dynamics of Surface Reactions and Oxygen Chemisorption

4.1	Introduction	50
4.2	Surface Reconstruction and "Oxide" Formation	52
4.3	Oxygen States at Metal Surfaces	55
4.4	Control of Oxygen States by Coadsorbates	64
4.5	Adsorbate Interactions, Mobility and Residence Times	65
4.6	Atom-tracking STM	69
4.7	Hot Oxygen Adatoms: How are they Formed?	71
4.8	Summary	72
	References	74
	Further Reading	75

Chapter 5 Catalytic Oxidation at Metal Surfaces: Atom Resolved Evidence

5.1	Introduction	77
5.2	Ammonia Oxidation	78
	5.2.1 Cu(110) Pre-exposed to Oxygen	79
	5.2.2 Coadsorption of Ammonia–Oxygen Mixtures at Cu(110)	81
	5.2.3 Coadsorption of Ammonia–Oxygen Mixtures at Mg(0001)	83
	5.2.4 Ni(110) Pre-exposed to Oxygen	83
	5.2.5 Ag(110) Pre-exposed to Oxygen	84
5.3	Oxidation of Carbon Monoxide	85
5.4	Oxidation of Hydrogen	89
5.5	Oxidation of Hydrocarbons	91
5.6	Oxidation of Hydrogen Sulfide and Sulfur Dioxide	95
5.7	Theoretical Analysis of Activation by Oxygen	98
5.8	Summary	99
References		100
Further Reading		102

Chapter 6 Surface Modification by Alkali Metals

6.1	Introduction	103
6.2	Infrared Studies of CO at Cu(110)–Cs	105
6.3	Structural Studies of the Alkali Metal-modified Cu(110) Surface	105
	6.3.1 Low-energy Electron Diffraction	105
	6.3.2 Scanning Tunnelling Microscopy	106
	6.3.3 Cu(110)–Cs System	107
	6.3.4 Oxygen Chemisorption at Cu(110)–Cs	108
6.4	Reactivity of Cu(110)–Cs to NH_3 and CO_2	111
6.5	Au(110)–K System	113
6.6	Cu(100)–Li System	115
6.7	Summary	117
References		119
Further Reading		120

Chapter 7 STM at High Pressure

7.1	Introduction	121
7.2	Catalysis and Chemisorption at Metals at High Pressure	123
	7.2.1 Carbon Monoxide and Nitric Oxide	124
	7.2.2 Hydrogenation of Olefins	126

7.3	Restructuring of the Pt(110)–(1 × 2) Surface by Carbon Monoxide	129
7.4	Adsorption-induced Step Formation	131
7.5	Gold Particles at FeO(111)	131
7.6	Hydrogen–Deuterium Exchange and Surface Poisoning	132
7.7	Summary	133
References		134
Further Reading		134

Chapter 8 Molecular and Dissociated States of Molecules: Biphasic Systems

8.1	Introduction	135
8.2	Nitric Oxide	136
8.3	Nitrogen Adatoms: Surface Structure	142
8.4	Carbon Monoxide	143
8.5	Hydrogen	145
8.6	Dissociative Chemisorption of HCl at Cu(110)	147
8.7	Chlorobenzene	148
8.8	Hydrocarbon Dissociation: Carbide Formation	150
8.9	Dissociative Chemisorption of Phenyl Iodide	150
8.10	Chemisorption and Trimerisation of Acetylene at Pd(111)	151
8.11	Summary	152
References		153
Further Reading		155

Chapter 9 Nanoparticles and Chemical Reactivity

9.1	Introduction	156
9.2	Controlling Cluster Size on Surfaces	157
9.3	Alloy Ensembles	159
9.4	Nanoclusters at Oxide Surfaces	160
9.5	Oxidation and Polymerisation at Pd Atoms Deposited on MgO Surfaces	165
9.6	Clusters in Nanocatalysis	167
9.7	Molybdenum Sulfide Nanoclusters and Catalytic Hydrodesulfurisation Reaction Pathways	169
9.8	Nanoparticle Geometry at Oxide-supported Metal Catalysts	171
9.9	Summary	175

References		176
Further Reading		178

Chapter 10 Studies of Sulfur and Thiols at Metal Surfaces

10.1	Introduction		179
10.2	Studies of Atomic Sulfur Adsorbed at Metal Surfaces		180
	10.2.1	Copper	181
	10.2.2	Nickel	185
	10.2.3	Gold and Silver	189
	10.2.4	Platinum, Rhodium, Ruthenium and Rhenium	190
	10.2.5	Alloy Systems	193
10.3	Sulfur-containing Molecules		195
10.4	Summary		199
References			200
Further Reading			202

Chapter 11 Surface Engineering at the Nanoscale

11.1	Introduction		203
11.2	"Bottom-up" Surface Engineering		204
	11.2.1	Van der Waals Forces	205
	11.2.2	Hydrogen Bonding	207
	11.2.3	Chiral Surfaces from Prochiral Adsorbates	208
	11.2.4	Covalently Bonded Systems	209
11.3	Surface Engineering Using Diblock Copolymer Templates		211
11.4	Summary		214
References			214
Further Reading			216

Epilogue   217

Subject Index   219

Abbreviations

AES	Auger Electron Spectroscopy
AFM	Atomic Force Microscopy
DFT	Density Function Theory
EELS	Electron Energy Loss Spectroscopy
ESCA	Electron Spectroscopy for Chemical Analysis
ESR	Electron Spin Resonance
EXAFS	Extended X-ray Absorption Fine Structure
FEEM	Field Electron Emission Microscopy
FIM	Field Ion Microscopy
FTIR	Fourier-Transform Infrared
HAADF	High-Angle Annular Dark Field
HREELS	High-resolution Electron Energy Loss Spectroscopy
HRTEM	High-resolution Transmission Electron Microscopy
IETS	Inelastic Electron Tunnelling Spectroscopy
IRAS	Infrared Reflection Absorption Spectroscopy
L	Langmuir (exposure of 1 Torr s)
LEED	Low-energy Electron Diffraction
ML	Monolayer
NEXAFS	Near Edge X-ray Absorption Fine Structure
PIS	Penning Ionisation Spectroscopy
RAIRS	Reflection Absorption Infrared Spectroscopy
SEM	Scanning Electron Microscopy
SEXAFS	Surface Extended X-ray Absorption Fine Structure
SNOM	Scanning Near-field Optical Microscopy
SPM	Scanning Probe Microscopy
STEM	Scanning Transmission Electron Microscopy
STM	Scanning Tunnelling Microscopy
STS	Scanning Tunnelling Spectroscopy
SXRD	Surface X-ray Diffraction
TPD	Temperature-programmed Desorption
TPR	Temperature-programmed Reaction
UHV	Ultra-high Vacuum
UPS	Ultraviolet Photoelectron Spectroscopy
XPS	X-ray Photoelectron Spectroscopy
XRD	X-ray Diffraction

Some Relevant Units – SI and Derived Units

Physical quantity	Name of unit	Symbol and definition
Length	metre	m
Length	ångstrom, nanometre	$1\,\text{Å} \equiv 10^{-10}\,\text{m} \equiv 0.1\,\text{nm}$ $1\,\text{nm} \equiv 10^{-9}\,\text{m} \equiv 10\,\text{Å}$
Time	second	s
Electric current	ampere	A
Thermodynamic temperature	kelvin	K
Frequency	hertz	$\text{Hz} \equiv \text{s}^{-1}$
Energy	calorie	$1\,\text{cal} = 4.184\,\text{J}$

Pressure Conversion Factors

1 atm	101325 Pa
1 atm	1.01325 bar
1 bar	10^5 Pa
1 mbar	10^2 Pa
1 Torr	1.332 mbar
1 Torr	133.32 Pa

Gas Exposure   1 L (langmuir) $= 1 \times 10^{-6}$ Torr s

CHAPTER 1
Some Milestones in the Development of Surface Chemistry and Catalysis

"To understand science it is necessary to know its history"

Auguste Comte

1.1 Introduction

If we are to appreciate the significance and implications for surface chemistry and catalysis of the emergence of scanning tunnelling microscopy (STM) over the last 15 years, it is important that we examine first the stepwise development of the subject that led to the present fundamental scientific base of current thinking. The interpretation of atom resolved evidence in surface-catalysed reactions will clearly rely on whether it provides confirmation of accepted mechanistic models or how these models have to be modified to take on board the new experimental data. It is in this context that we view the development of STM as a significant step forward in the fundamental understanding of solid surfaces and their chemical reactivity. At the Nobel Symposium held in Sweden in 1978, J.S. Anderson presented a lecture entitled "Direct imaging of atoms in crystals and molecules", where he emphasised[1] how high-resolution electron microscopy should provide information on *local structure* in solids as distinct from *averaged crystal structures*, and therefore significant for the understanding of disordered solids, defects and non-stoichiometry. With a resolution of 2.5 Å, Anderson emphasised how chemists could benefit in being able to resolve the problem of how to relate structure and reactivity of disordered solids – including catalysts. The problem was even more severe for those interested in *surface reactivity*, and this is where STM had a major role to play. Low-energy diffraction had provided a breakthrough in the structural analysis of surfaces but its insensitivity to local disorder was a disadvantage when relating chemical reactivity to specific structural sites. It is instructive to consider briefly how the subject has developed over the last 80 years and prior to the emergence of STM in surface chemistry in 1990 by examining what was topical every few years and evident in the scientific literature of that time. The choice of article is a subjective one but may be helpful for those new to surface catalysis to obtain a view on the milestones in its development, particularly from the academic viewpoint.

1.2   1926: Catalysis Theory and Practice; Rideal and Taylor[2]

The dominant theme is the emergence of adsorption isotherms as an approach to relating gas pressure to the adsorbed state, with the solid being represented as a "latticework" of fixed atoms, the process of adsorption being viewed as equilibrium between two distinct processes, condensation and evaporation. Provided that the process is reversible, then it could be treated thermodynamically. Implicit in this is that molecules may reside at the surface for "some time" – what we will discuss later as the "residence time" – before desorbing. The concept of the "unimolecular layer" of adsorption was emphasised and its relation to gas pressure described by the mathematical form of the various isotherms – Freundlich, Langmuir and Polanyi. Kinetic studies of the adsorption process became significant with evidence for the dissociation of hydrogen at a tungsten surface obeying a square-root dependence on the pressure, $p^{1/2}$. Supporting this was the experimental evidence that the desorption process conformed to a second-order process arising from to the recombination of hydrogen adatoms.

1.3   1932: Adsorption of Gases by Solids; Faraday Discussion, Oxford[3]

There is further emphasis on adsorption isotherms, the nature of the adsorption process, with measurements of heats of adsorption providing evidence for different adsorption processes – physical adsorption and activated adsorption – and surface mobility. We see the emergence of physics-based experimental methods for the study of adsorption, with Becker at Bell Telephone Laboratories applying thermionic emission methods and work function changes for alkali metal adsorption on tungsten.

1.4   1940: Seventeenth Faraday Lecture; Langmuir[4]

It was usually assumed that the $(1 - \theta)$ factor in the Langmuir equation, $bP = \theta/(1 - \theta)$, took account of the fraction of the surface that was bare and that therefore the fraction of atoms (e.g. caesium on tungsten) that condense at the surface is proportional to $(1 - \theta)$. Langmuir in his lecture (given in 1938) drew attention to the physical assumptions underlying this factor $(1 - \theta)$ being very improbable, referring specifically to his experiments concerning the adsorption of caesium vapour on tungsten: when the coverage was close to unity, all the caesium atoms impinging on the surface were adsorbed, indicating that they sought vacant sites on the surface and were mobile.

Langmuir made the point in his lecture that the "lifetime τ of an atom at the surface is not independent of the presence of other atoms, being given by $\tau = \tau_0(1 - \theta)$. The shortening of the lifetime τ as θ approaches unity is the result of strong repulsive forces between pairs of atoms which occupy single sites". We will see that this view is central to what STM revealed some 60 years later.

1.5   1950: Heterogeneous Catalysis; Faraday Discussion, Liverpool[5]

Although there was the realisation that "clean" metal surfaces were essential to progress the understanding of adsorption and catalysis, it was J. K. Roberts and Otto Beeck who, as experimentalists, moved the subject forward, with Roberts' studies of hydrogen and oxygen adsorption at tungsten wires, cleaned by flashing to 2000 °C, and Beeck using large surface area metal films. Roberts had introduced earlier the distinction between immobile and mobile adsorption on fixed or localised sites with Miller discussing how statistical mechanics could be used to examine the equilibrium distribution in the mobile state and how it is related to the experimentally observed variation in the heat of adsorption with surface coverage.

Beeck at Shell Laboratories in Emeryville, USA, had in 1940 studied chemisorption and catalysis at polycrystalline and "gas-induced" (110) oriented porous nickel films with ethene hydrogenation found to be 10 times more active than at polycrystalline surfaces. It was one of the first experiments to establish the existence of structural specificity of metal surfaces in catalysis. Eley suggested that good agreement with experiment could be obtained for heats of chemisorption on metals by assuming that the bonds are covalent and that Pauling's equation is applicable to the process $2M + H_2 \rightarrow 2M-H$.

Lennard-Jones in the Introduction to his paper stated "*The literature pertaining to the sorption of gases by solids is now so vast that it is impossible for any, except those who are specialists in the experimental details, to appreciate the work which has been done or to understand the main theoretical problems which require elucidation*". He goes on to describe what is still one of the cornerstones of adsorption behaviour, the Lennard-Jones potential energy diagram, its explanation of "activated adsorption" and its relevance as an important concept in the understanding of surface catalysis. The paper by Volmer considers experimental evidence for the migration of molecules at surfaces from the viewpoint of crystal growth. He emphasises the need to search for experimental evidence for "two-dimensional mobility" and discusses Estermann's data for silver on quartz and benzophenone on mica surfaces.

What was evident in 1950 was that very few surface-sensitive experimental methods had been brought to bear on the question of chemisorption and catalysis at metal surfaces. However, at this meeting, Mignolet reported data for changes in work function, also referred to as surface potential, during gas adsorption with a distinction made between Van der Waals (physical) adsorption and chemisorption. In the former the work function decreased (a positive surface potential) whereas in the latter it increased (a negative surface potential), thus providing direct evidence for the electric double layer associated with the adsorbate.

The work of Beeck and Roberts had a strong influence on the need to characterise the chemical state of metal surfaces under different preparative conditions, *i.e.* whether it was a metal filament, a high area metal film or a catalyst formed by the reduction of a metal oxide. Wheeler in 1952 highlighted the potential conflict between the "clean" surface and "bulk catalyst" approaches to catalyst research.[6] There was emerging a driving force to develop experimental methods which relied on ultra-high vacuum techniques, where the background pressure was 10^{-9} mbar or less, as prerequisites for studies of chemisorption and chemical reactivity of metal surfaces. In 1953, one of us (M.W.R.) attended a Summer School on "The Solid State and Heterogeneous Catalysis" at the University of Bristol, "intended for those engaged in research in University, Government and Industrial Laboratories". This consolidated the messages that had emerged from the Faraday meeting of 1950, with Stone and Gray emphasising the defect solid state and Eley drawing attention to the problems associated with metal surfaces prepared by various methods. Mobility of surface atoms was anticipated to occur when the temperature of the solid was above $0.3T_m$, whereas mobility of the bulk atoms occurred above $0.5T_m$ (the Tammann temperature), where T_m is the melting point in kelvin of the solid. In contrast to what we shall discuss later, surface mobility was considered to be a phenomenon to be associated with "high temperatures" and therefore in accord with Langmuir's concept of the checkerboard model of a surface being homogeneous and consisting of fixed surface sites!

1.6 1954: Properties of Surfaces[7]

This conference, organised by the New York Academy of Sciences, emphasised the contribution that fundamental studies carried out in industry were making, papers emanating from Bell Telephone Laboratories, Westinghouse Research Laboratories, du Pont Nemours, Kodak, Sylvania Electric Products and General Electric (the "home" of Irving Langmuir). Although there was much emphasis given to the physics of surfaces, we draw attention to two papers, the first by Becker and the second by H.A. Taylor. It is clear that Becker was greatly influenced by the development of the field emission microscope and what it revealed about "foreign atoms" adsorbed on metal surfaces and how the work function varies from one crystal face to another, and that "in adsorption the arrangement of the surface metal atoms plays an important part". Becker refers to the possibility of measuring sticking probabilities using the "flash filament" (later called temperature-programmed desorption TPD) technique with the emergence of ultra-high vacuum techniques and the ion gauge for pressure measurement. His paper emphasises how these developments led him to reappraise his article "The life history of adsorbed atoms and ions" published in 1929.[8]

H.A. Taylor considers kinetic aspects of surface reactions and starts from the proposition that although in discussions of reaction kinetics it is customary to divide the subject into two classes, homogeneous and heterogeneous, the inference that what may be true for one class cannot be true for the other. Taylor took the view that a single basis must underlie both classes of reactions, that each must be governed by the same basic principles and that no chasm exists between them. A paper with Thon questions the checkerboard model and that the surface plays an active rather than a passive role as implied in the Langmuir model. He questions the use of orders of reaction as providing unambiguous models of surface reactions. Taylor was particularly attracted to the views of the Russian scientist Semenov, who regarded the solid surface in a catalytic reaction as a source for generating and terminating radical reactions. The Taylor–Thon view led to the rejection of mechanisms based on the reaction between two chemisorbed species and favoured the reaction between a chemisorbed species and a gaseous reactant (essentially an Eley–Rideal mechanism). The Semenov view was that even in a heterogeneously catalysed reaction the product was formed by the reaction of a free radical and an "inert molecule", just as in a homogeneous chain reaction.

Can STM throw light on whether homogeneous gas-phase and heterogeneous surface reactions encompass a common theme – the participants of surface radicals in a "two-dimensional gas"?

1.7   1957: Advances in Catalysis; International Congress on Catalysis, Philadelphia[9]

The first International Congress on Catalysis to be held in America was in Philadelphia in 1956 and according to Farkas it was *"in view of the tremendous growth in the industrial applications of catalysis and the ever increasing scientific activity in the field"*. There was at this meeting an obvious step-change in the science, with new experimental methods being introduced to investigate solid surfaces and their chemical reactivity. In particular, there was the emergence of low-energy electron diffraction (LEED) (Schlier and Farnsworth), infrared studies (Eischens and Pliskin), magnetic studies (Selwood), isotopic exchange studies (Bond and Kemball), electron spin resonance (Turkevich), conductivity studies (Suhrmann) and flash-filament, later renamed temperature-programmed desorption (Ehrlich). Surface cleanliness of metal surfaces had become a fundamental issue and a pointer to the development of what became referred to as the surface science approach to catalysis.

1.8 1963: Conference on Clean Surfaces with Supplement: Surface Phenomena in Semiconductors, New York[10]

This was an outstanding meeting held in New York, which to at least to one of us (M.W.R.) marked a turning point in surface science. In the Panel Discussion, R.S. Hansen, a chemist, made the telling comment: *"The semiconductor physicist is encountering a number of chemical problems that he is not trained to solve; the chemist on the other hand is unaware of these potentially very interesting problems. I can say from a chemist's viewpoint I am sure that part of this difficulty is the language barrier between the physicist and the chemist and that certain of the concepts of the physicist are stated in language, where he is very much at home, that are purely phenomenological and have no strictly scientific context"*. Hansen goes on to give as an example "slow" or "fast" surface states in explaining conductivity changes in semiconductors. There were some outstanding papers which set the scene for the development of surface science: field emission (Müller); slow electron diffraction (Germer, also Farnsworth); work function and photoelectric measurements (Gobeli and Allen); adsorption at clean surfaces (Ehrlich); reactions of hydrocarbons with clean rhodium surfaces (R.W. Roberts); nucleation of adsorbed oxygen on clean surfaces (Rhodin); dynamic measurements of adsorption of gases on clean tungsten surfaces (Ricca); and oxygen complexes on semiconductor surfaces (Mino Green).

1.9 1966: Faraday Discussion Meeting, Liverpool[11]

The significance and impact of surface science were now becoming very apparent with studies of single crystals (Ehrlich and Gomer), field emission microscopy (Sachtler and Duell), calorimetric studies (Brennan and Wedler) and work function and photoemission studies (M.W.R.). Distinct adsorption states of nitrogen at tungsten surfaces (Ehrlich), the facile nature of surface reconstruction (Muller) and the defective nature of the chemisorbed oxygen overlayer at nickel surfaces (M.W.R.) were topics discussed.

1.10 1967: The Emergence of Photoelectron Spectroscopy[12]

Siegbahn's publication of his group's development in Uppsala of what was described as ESCA (electron spectroscopy for chemical analysis) opened up the field of photoelectron spectroscopy, which through an understanding and interpretation of shifts in binding energy provided much more than the acronym suggested – chemical analysis. It is interesting to recall his comment regarding core-level binding energies: *"We had discovered what we call the chemical shift. In the beginning we didn't like this: we were physicists and wanted to study systematically the behaviour of elements. There was now a problem as we had to be careful that the substance was not oxidized or changed in some other way"*. Kai Siegbahn was awarded the Nobel Prize for Physics in 1981. The surface chemist took advantage of the chemical shift in being able to distinguish different bonding states of the same element, for example N(a), NH(a) and NH_3(a) and differentiating between molecular and dissociated states which had previously relied on whether first-order (molecular state) or second-order (dissociated state) desorption kinetics were observed.

1.11 1968: Berkeley Meeting: Structure and Chemistry of Solid Surfaces[13]

This meeting was organised by Gabor Somorjai, driven by the rapid development of experimental methods in what was now developing as a sub-set of heterogeneous catalysis – surface science. It was evident that over the 2 years since the Faraday Discussion Meeting in 1966 the subject had moved on apace, with low-energy electron diffraction (LEED) following Germer's work being a dominant theme. Auger electron spectroscopy had just come into prominence with Weber and Peria, following Harris at General Electric's laboratories at Schenectady, realising that the LEED equipment could be easily adapted to enable Auger spectra to be obtained, which provided chemical analysis of the surface. There was an overwhelming emphasis on studies of single crystals.

A number of papers stand out. May and Germer, using LEED, investigated the interaction of hydrogen with chemisorbed oxygen at Ni(110) at 450 K. They recognised the presence of (2×1) oxygen islands and that their attack by hydrogen "was assumed to be effective along the island's perimeters". With the emergence of STM, some 40 years later, this model is seen to hold at the atom resolved level. Tracey and Blakely drew attention to the limitations of LEED (at that time), which, although providing evidence on the symmetry of surface structures, did not define either the surface coverage or the precise atomic arrangement, an aspect that was pursued subsequently with much vigour.

1.12 1972: A Discussion on the Physics and Chemistry of Surfaces, London[14]

This meeting was held at The Royal Society, London, and was organised by J.W. Linnett. There were 11 papers with theoretical inputs but with more emphasis given to new developments in experimental methods including structural (LEED and electron microscopy) and surface spectroscopies. LEED provided crucial evidence for the role of surface steps at platinum single crystals in the dissociation of various diatomic molecules, while electron microscopy revealed the role of dislocations as sites of high reactivity of graphite. The advantage of coupling Auger spectroscopy with LEED, the submonolayer sensitivity of XPS, the emergence of RAIRS for studying chemisorption at single-crystal metal surfaces and the reactivity of structural defects stood out among the papers.

There was now the possibility of obtaining quantitative data for the chemical state of solid surfaces and reactions at single-crystal surfaces, provided as a prerequisite that the spectroscopy (AES and XPS) was coupled with the rigours associated with ultra-high vacuum techniques, which ensured that surface contamination was negligible. This was achieved through, for example, the development in 1971 by Vacuum Generators at East Grinstead, UK, of the ESCA-3 UHV spectrometer with multiphoton (X-ray and UV) sources. There was a step-change in experimental strategies for investigating chemisorption and catalytic reactions at metal surfaces and almost immediately there were significant advances, the science moving from qualitative logic to quantitative understanding with hitherto unavailable structural and chemical information.

1.13   1987: Faraday Symposium, Bath[15]

The implications of the role of non-thermalised oxygen transients in oxidation catalysis and the limitations of the classical kinetic approach based on the Langmuir–Hinshelwood and Eley–Rideal mechanisms were first discussed. The basis for these was spectroscopic (XPS) data for the coadsorption of ammonia and oxygen at Mg(0001) surfaces where $O^{\delta-}(s)$ transients were the active oxidants. The general significance of kinetically hot mobile transients in oxidation catalysis at metal surfaces was established, but with some scepticism being expressed by some concerning the concept. However, in 1992, Ertl's STM data for oxygen chemisorption at aluminium surfaces coupled with earlier oxidation studies of carbon monoxide by XPS provided convincing evidence for the concept to receive much wider acceptance, particularly through combining STM with chemical information from XPS. Although Binnig and Rohrer had made an impact with the development of STM in 1982, there was no serious application reported in surface catalysis until the early 1990s.

In 1991, Güntherodt and Wiesendanger edited *Scanning Tunneling Microscopy I* and in 1994 the second edition was published.[16] In the Preface to the second edition, Wiesendanger drew attention to the progress made in the application of STM "most notably in the field of adsorbates and molecular systems"; Wintterlin, Behm and Chiang contributed with examples of oxygen chemisorption, alkali metal adsorption and the molecular imaging of organic molecules in an additional chapter, "Recent Developments".

1.14   Summary

As suggested by Thomas in his 1994 article "Turning Points in Catalysis",[17] Europe was "*the crucible and fulcrum for change in the science and technology* for more than two centuries but has developed to such an extent over the last 50 years that the pursuit of catalysis transcends both national and continental boundaries". The Kendall Awards of the American Chemical Society in Colloid and Surface Chemistry initiated in 1954 illustrate how catalysis and surface science became dominant in the USA and Canada in the 1970s and 1980s, with awards being made to Burwell (1973), Keith Hall (1974), Gomer (1975), Boudart (1977), Somorjai (1981), Ehrich (1982), Ruckenstein (1986), Yates (1987) and White (1990). In contrast, during the earlier period 1954–1970, catalysis was very much less prominent amongst the Awards,[18] more emphasis being given to colloid chemistry and liquid surfaces.

In this chapter, we have chosen from the scientific literature accounts of symposia published at intervals during the period 1920–1990. They are personal choices illustrating what we believe reflect significant developments in experimental techniques and concepts during this time. Initially there was a dependence on gas-phase pressure measurements and the construction of adsorption isotherms, followed by the development of mass spectrometry for gas analysis, surface spectroscopies with infrared spectroscopy dominant, but soon to be followed by Auger and photoelectron spectroscopy, field emission, field ionisation and diffraction methods.

Although Langmuir's checkerboard model is still valid and retained in models of adsorption and surface reactivity, there was experimental evidence for two-dimensional mobility of the adsorbate, with Volmer[3] at a Faraday Discussion as early as 1932 discussing a number of different systems including silver and benzophenone on quartz. Semenov took a view that the catalyst surface was an "agent" generating radicals and drew attention to a possible analogy with homogeneous gas-phase reactions.

The Royal Society of Chemistry in 2003 published a special volume to mark the centenary of the founding of the Faraday Society, which consisted of 23 papers reprinted from Faraday journals.[19] Each article was selected by a scientist active in a particular field but requested by the President, Ian Smith, to add a commentary emphasising the impact that the chosen paper had on their own work. In the field of catalysis, J.M. Thomas chose the paper by Rabo *et al.* on "Studies of Cations in Zeolites", which had a profound impact on his work in the early 1980s as it gave authoritative accounts of structural aspects, adsorbability and reactivity of cation-exchanged zeolites. It was a paper by H.S. Taylor, the 5th Spiers Memorial Lecture published in 1950, that one of us chose in the field of gas–solid surface science as it emphasised the need to bring together two rather conflicting philosophies of tackling the fundamentals of heterogeneous catalysis: the clean surface and bulk catalyst approaches. Taylor (with E.K. Rideal) were pioneers in the development of the subject since the early part of the 20th century and made the following remark to close his lecture:

> "We may anticipate a reconciliation of the several attitudes that sometimes have appeared to divide us, but are in reality a spur to further and continued effort towards the mastery of our science in an era which is of deep

significance in all human affairs. In prospect, therefore, the future of our science is both challenging and bright".

It is a remark that is appropriate even 50 years later!

The development of experimental methods over the last 50 years has been at the forefront of new strategies that emerged, driven by the need to obtain molecular information relevant to the structure of catalyst surfaces and the dynamics of surface reactions. The ultimate aim was in sight with the atomic resolution that became available from STM, particularly when this was coupled with chemical information from surface-sensitive spectroscopies.

Although tunnelling spectroscopy was first applied by Giaver[20] in superconductivity in 1961, its application in the form of the scanning tunnelling microscope was not described by Binnig and Rohrer[21] until 1983. In this paper, the authors state *"These initial results demonstrate that STM shows a great potential for surface studies. Even more, the possibility of determining work functions and for performing tunnelling spectroscopy with atomic resolutions should make vacuum tunnelling a powerful technique for solid state physics and other areas".* One of the IBM team that met at the Research Conference in Oberlach in 1986, Christoph Gerber, when interviewed[22] recently, said *"When the reconstruction of the silicon 7 × 7 was recorded on the 2D plotter atom by*

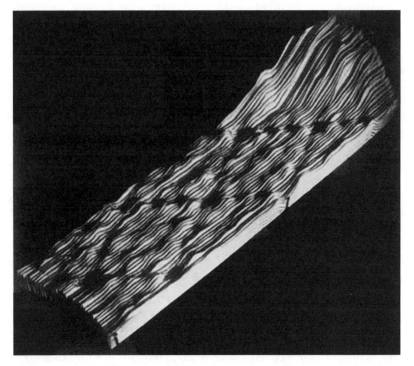

Figure 1.1 The first STM image of the 7 × 7 reconstruction of Si (111) assembled from the original recorder traces of Binnig *et al.* (Reproduced from Ref. 23).

atom (very slowly) everybody knew that something very special had been witnessed...". (Figure 1.1; see also Chapter 3). It is on surface chemistry and catalysis that we focus in this book.

References

1. J. S. Anderson, *Chem. Scr.*, 1979, **14**, 287.
2. H. S. Taylor and E. K. Rideal, *Catalysis in Theory and Practice*, Macmillan, London, 1926.
3. Adsorption of Gases by Solids, General Discussion, Faraday Society, London, 1932.
4. I. Langmuir (Seventeenth Faraday Lecture), *J. Chem. Soc.*, 1940, 511.
5. Heterogeneous Catalysis, *Discuss. Faraday Soc.*, 1950, No. 8.
6. A. Wheeler, in: *Structure and Properties of Solid Surfaces*, ed. R. Gomer and C. S. Smith, University of Chicago Press, Chicago, 1953, 439.
7. W. Miner (ed.), *Properties of Surfaces Ann. N. Y. Acad. Sci.*, 1954, **58**.
8. J. A. Becker, *Trans. Electrochem Soc.*, 1929, **55**, 153.
9. A. Farkas (ed.), *Proceedings of International Congress on Catalysis, Advances in Catalysis*, Academic Press, New York, 1957.
10. M. C. Johnstone (ed.), *Conference on Clean Surfaces with Supplement: Surface Phenomena in Semiconductors Ann. N. Y. Acad. Sci.*, 1963, **101**.
11. The Role of the Adsorbed State in Heterogeneous Catalysis, *Discuss. Faraday Soc.*, 1966, No. 41.
12. K. Siegbahn, C. Nordling, A. Fahlman, R. Nordberg, K. Hamrin, J. Hedman, G. Johansson, T. Bergmark, S. E. Karlsson, I. Lindgren and B. Lindberg, E.s.c.a.: atomic, molecular and solid state structure studied by means of electron spectroscopy, *Nova Acta Soc. Sci. Uppsala, Ser. IV*, 1967, **20**.
13. G. A. Somorjai, (ed.), *Structure and Chemistry of Solid Surfaces*, Wiley Inc., New York, 1969.
14. Discussion of the Physics and Chemistry of Surfaces, *Proc. R. Soc. Lond., Ser. A*, 1972, **331**.
15. Promotion in Heterogeneous Catalysis, *J. Chem. Soc., Faraday Trans. 1*, 1987, Faraday Symposium 21.
16. H.-J. Guntherodt and R. Wiesendanger (eds.), *Scanning Tunneling Microscopy I*, Springer, Berlin, 1994.
17. J. M. Thomas, *Angew. Chem. Int. Ed.*, 1994, **33**, 913.
18. T. Fort and K. J. Mysels (eds), *Eighteen Years of Colloid and Surface Chemistry The Kendall Award Addresses 1973–1990*, American Chemical Society, Washington DC, 1991.
19. *100 Years of Physical Chemistry, a Celebration of the Faraday Society*, Royal Society of Chemistry, Cambridge, 2003.
20. I. Giaver, *Phys. Rev. Lett.*, 1961, **5**, 147.
21. G. Binnig and H. Rohrer, *Surf. Sci.*, 1983, **126**, 236.
22. Omichron News, *Pico*, 2006, **10**, No. 1.
23. G. Binnig, H. Rohrer, C. Gerber and E. Weibel, *Phys. Rev. Lett.*, 1983, **50**, 120.

Further Reading

C. B. Duke (ed.), *Surface Science: The First Thirty Years*, North-Holland, Amsterdam, 1994.

M. W. Roberts, Heterogeneous catalysis since Berzelius: some personal reflections, *Catal. Lett.*, 2000, **67**, 1.

C. B. Duke and E. W. Plummer (eds), *Frontiers in Surface and Interface Science*, Elsevier, Amsterdam, 2002.

R. L. Burwell, Jr, A retrospective view of advances in heterogeneous catalysis: 1956–1996, science, in *11th International Congress on Catalysis – 40th Anniversary*, Elsevier, Amsterdam, 1996.

H. Heinemann, A retrospective view of advances in heterogeneous catalysis: 1956–1996, technology, in *11th International Congress on Catalysis – 40th Anniversary*, Elsevier, Amsterdam, 1996.

G. Suits (ed.), *The Collected Works of Irving Langmuir, Surface Phenomena*, Vol. 9, Pergamon Press, Oxford, 1962.

CHAPTER 2
Experimental Methods in Surface Science Relevant to STM

There was an electron in gold
Who said "Shall I do what I'm told?
Shall I snuggle down tight
With a brief flash of light
Or be Auger outside in the cold?"

Arthur H. Snell

2.1 Introduction

Experimental methods in surface science are considered briefly in order to illustrate how experimental data and concepts that emerged from their application could be progressed through evidence from STM at the atom resolved level. They include kinetic, structural, spectroscopic and work function studies. Further details of how these methods provided the experimental data on which much of our present understanding of surfaces and their reactivity can be obtained from other publications listed under Further Reading at the end of this chapter.

2.2 Kinetic Methods

The classical approach for discussing adsorption states was through Lennard-Jones potential energy diagrams and for their desorption through the application of transition state theory. The essential assumption of this is that the reactants follow a potential energy surface where the products are separated from the reactants by a transition state. The concentration of the activated complex associated with the transition state is assumed to be in equilibrium with the gas-phase reactants. The latter can be considered in terms of partition functions – translational, vibrational and rotational – with the activated complex being thought of as having one loose vibrational mode that corresponds to the motion leading to products. Although this approach, following the concepts developed by Glasstone, Laidler and Eyring,[1] had many attractions in providing a physico-chemical process for a surface-catalysed reaction, it lacked the ability to provide a unique solution, particularly when it also included an arbitrary transmission coefficient, maximum value unity, which defines the probability that the activated complex passes along the reaction product channel. Whether one can assume that equilibrium exists between the reactants and the transition state complex is also an assumption that requires support from experimental scrutiny at the atom resolved level. Heats of adsorption were determined calorimetrically and activation energies of desorption, E_{des}, from kinetic studies, thermal desorption spectroscopy being widely used where the Polanyi–Wigner relationship:

$$-\frac{d\theta}{dt} = \nu_n \theta^n \exp\left(\frac{E_{des}}{RT}\right) \qquad (1)$$

is central to analysing the desorption–temperature relationship, where θ is the surface coverage, n is the order of the process and ν is the vibrational frequency. In general, two situations are observed with $n = 1.0$ or 2.0, with first-order desorption characterising molecular desorption and second-order desorption interpreted as the recombination of dissociative states. Although analysis of adsorption data via the assumptions associated with particular isotherms (Langmuir, Freundlich, *etc.*) were an essential aspect of the development of models of catalytic reactions, the participation of precursor-mediated adsorption was highlighted by Kisliuk,[2] who developed an appropriate kinetic model. Modifications of this have been described by Madix and co-workers[3] and King and Wells[4] to account for a combination of direct and precursor-mediated adsorption and the effect of lateral interactions. Adsorption studies at low temperatures (80 K) provided more direct experimental evidence for a precursor state, first by Ehrlich[5] and subsequently by others, including work function studies of nitrogen at tungsten surfaces.[6] We shall see that low-temperature studies have provided an essential stimulus in STM studies of the dynamics of surface reactivity. A detailed discussion of kinetic methods in elucidating the macroscopic nature of adsorption states is discussed widely elsewhere.[7]

2.3 Vibrational Spectroscopy

This has been one of the most significant experimental methods for obtaining structural information on adsorbed species. Initially through the studies of the Russian group led by Terenin in Leningrad and Eischens and Pliskin at Texaco using infrared transmission methods with metals supported on high surface area adsorbents (Al_2O_3, SiO_2), linear and bridge-bonded states of chemisorbed carbon monoxide were delineated.[8] Infrared methods were initially confined in the main to studies of carbon monoxide in view of the latter's high extinction coefficient, but was limited in that it could not confirm whether the molecule was dissociated. In the early 1970s, reflection absorption infrared spectroscopy (RAIRS) opened the way to the possibility of investigating adsorption at metal single-crystal surfaces following Greenler and Tompkins' theoretical work and Pritchard and colleagues' study of the copper–carbon monoxide system.[9] Sheppard has recently critically reviewed the contribution that vibrational spectroscopy has made in determining the structure of adsorbed states of carbon monoxide and hydrocarbons at metal surfaces,[10] while Somorjai's group has studied CO oxidation at high pressures using sum frequency generation (SFG), an experimental method that is surface specific, being insensitive to both the gas phase and the bulk solid. The principle of the SFG process is described briefly by Somorjai and Marsh,[11] but a more detailed description has been given by Shen.[12]

2.4 Work Function

Changes in the work function of surfaces by adsorbed species can be measured by a number of different experimental methods: photoelectric; field emission; the diode and vibrating capacitor methods. The last is the most versatile method, with the design proposed by Mignolet in 1950 enabling changes in work function to be followed over a wide range of gas pressures and not just confined to "vacuum" conditions as with the other methods.[13] A prerequisite for the success of the capacitor method was that the surface of the static reference electrode, usually gold, was not influenced by the gas being investigated. It was one of the first surface-sensitive techniques to be used in the study of chemisorption at metal surfaces and a good review of the field, as it emerged in the late 1950s, is that by Culver and Tompkins.[14] For hydrogen and oxygen chemisorbed at nickel surfaces, the work function of the metal was increased by 0.35 and 1.6 eV, respectively, whereas for a physically adsorbed state, such as Xe, the work function was reported to decrease by 0.85 eV. Interpretation of the observed values was dependent on the surface dipole and nature of the surface bond: van der Waals, ionic or covalent. Although molecules chemisorbed at metal surfaces were considered to be characterised by a particular surface potential – usually an increase in the work function of the metal – and which field emission microscopy showed was crystal plane specific, it became clear in the early 1960s that for oxygen there were competitive processes, one leading to an increase in the metal work function and the other to a decrease. In the case of nickel, it was proposed[15] that chemisorbed oxygen was metastable and at low temperatures the surface reconstructed with the formation of a defective oxide in the temperature range 80–295 K. This was further confirmed[15] by studying the photoelectron yield (Figure 2.1), the energy distribution of the emitted photoelectrons and later by X-ray photoelectron spectroscopy. The emergence of STM has enabled these views to be further addressed.

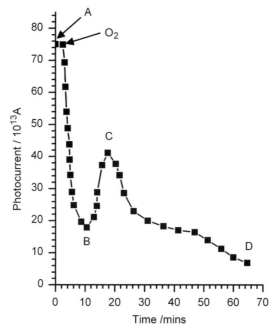

Figure 2.1 Real-time photoemission study ($h\nu = 6.2$ eV) of the interaction of oxygen ($P_{O_2} = 10^{-6}$ Torr) with a nickel surface at 300 K. The photocurrent decreases initially (A–B), then recovers (B–C), before finally decreasing (C–D). Surface reconstruction occurs (B–C) with further support from studies of the work function. The work function measured by the capacitor method[15] increases by 1.5 eV with oxygen exposure at 80 K followed by a rapid decrease on warming to 295 K and an increase on further oxygen exposure at 295 K. These observations suggest that three different oxygen states are involved in the formation of the chemisorbed overlayer. (Reproduced from Refs. 15, 42).

2.5 Structural Studies

In 1961, Germer and MacRae described, in their paper given under the auspices of the Robert A. Welch Foundation, "A new low-energy electron diffraction technique having possible application to catalysis" (see Further Reading). It was a significant advance from that adopted earlier by Farnsworth, in that the diffraction of low-energy electrons could be observed directly on a fluorescent screen, providing information on surface structure.[16]

Although in early LEED studies all new diffraction features, as for example for oxygen chemisorption at Ni(110), were attributed to diffraction by the "heavy" metal atoms, it became clear that "light" adsorbate atoms were also effective scatterers of low-energy electrons. LEED is, however, particularly sensitive to long-range order and disordered surface structures are either not revealed or are indicated by diffuse or streaked features (see Further Reading). The emphasis was, therefore, on reporting structures arising from well-defined, sharp diffraction features from which bond distances were calculated. The physics of the diffraction process and the interpretation of LEED patterns providing structural information have been considered extensively by Pendry[17] and others. It is in revealing short-range or disordered surface structures, however, that STM has a distinct advantage over LEED, with the possibility of relating such disordered states to reactivity and catalysis.

In the majority of cases where adsorbates form ordered structures, the unit cells of these structures are longer than that of the substrate; they are referred to as superlattices. Two notations are used to describe the superlattice, the Wood notation and a matrix notation.[18] Some examples of overlayer structures at an fcc(110) surface are as follows:

$$
\begin{array}{cc}
\text{Wood notation} & \text{Matrix notation} \\
p(2 \times 1) & \begin{pmatrix} 2 & 0 \\ 0 & 1 \end{pmatrix} \\
p(3 \times 1) & \begin{pmatrix} 3 & 0 \\ 0 & 1 \end{pmatrix} \\
c(2 \times 2) & \begin{pmatrix} 1 & -1 \\ 1 & 1 \end{pmatrix}
\end{array}
$$

The prefixes c and p mean "centred" and "primitive", respectively, where centred refers to when an adsorbate is added in the centre of the primitive unit cell.

In an attempt to relate defective LEED patterns observed during the chemisorption of oxygen at Cu(210), an optical simulation based on some simple models of chemisorption, in which oxygen dissociation at a limited number of surface sites was followed by extensive surface diffusion prior to coming to rest, provided the best matrix for simulating the observed LEED pattern.[19] It is interesting particularly with what is now known from STM that elongated oxygen structures provided the most suitable matrix model.

Some conclusions that emerged in 1978 from the optical simulation study were as follows; these could only be tested by (future) STM studies:

(a) No non-nucleation model could account for the LEED patterns observed.
(b) Correlated or semi-correlated diffusion by a hopping (surface diffusion) process over substantial distances (~ 10 nm) is required to account for the observations.
(c) The nucleation model requires considerable diffusion of the molecular precursor state to account for the high initial sticking probability. At the ultimate coverage attained, the adlayer contains phase boundaries and vacancies which give rise to streaked half-order spots.

Weinberg and co-workers' study in 1982[20] of the chemisorption of carbon monoxide on Ru(0001) by LEED exemplifies how lateral interactions determine the periodicity of the ordered overlayer and island structures. The formation of small islands is attributed to limited diffusion of the molecules on the surface, with surface steps playing an important role in the growth of islands. We shall see how these concepts relate to what STM revealed at the atom resolved level.

Heinz and Hammer[21] have recently provided a convincing and elegant appraisal of the limitations of both LEED and STM when applied as separate tools for structural studies, but that when combined provide a powerful approach. It is shown that in most cases STM images provide the key to the applicable structural model and are invaluable when a large number of different models need to be tested. The authors argue that the combination of STM and quantitative LEED provides a new, powerful approach for access to complex surface structures. They also make the observation that this makes a significant demand on the UHV equipment necessary, the availability of experimental expertise with both techniques available in the same laboratory and complemented by a theoretical group for intensity analysis. It is perhaps not surprising that the ideal combination of STM and LEED is not widely implemented!

Local surface structure and coordination numbers of neighbouring atoms can be extracted from the analysis of extended X-ray absorption fine structures (EXAFS). The essential feature of the method[22] is the excitation of a core-hole by monoenergetic photons; modulation of the absorption cross-section with energy above the excitation threshold provides information on the distances between neighbouring atoms. A more surface-sensitive version (SEXAFS) monitors the photoemitted or Auger electrons, where the electron escape depth is small (~ 1 nm) and discriminates in favour of surface atoms over those within the bulk solid. Model compounds, where bond distances and atomic environments are known, are required as standards.

2.6 Photoelectron Spectroscopy

It was Kai Siegbahn, Nobel Laureate in 1981, who pioneered the application of X-ray photoelectron spectroscopy (XPS) in physics and chemistry.[23] This was through relating the binding energy E_B of electrons in orbitals with the charge associated with the atom involved in the photoemission process. The experiment involves the measurement of the kinetic energy E_K of the electron emitted by a photon of energy hv and through the principle of conservation of energy [eqn. (2)] calculating the binding energy E_B, which through Koopmans theorem can be equated to the orbital energy:

$$E_K = hv - E_B \qquad (2)$$

Accompanying the photoemission process, electron reorganisation can result in the ejection of a photon (X-ray fluorescence) or internal electronic reorganisation leading to the ejection of a second electron. The latter is referred to as the Auger process and is the basis of Auger electron spectroscopy (AES). It was Harris at General Electric's laboratories at Schenectady, USA, who first realised that a conventional LEED experiment could be modified easily to provide Auger spectra using the hot electron-emitting filament for generating "holes" in the electronic structure of the substrate which initiated the Auger process. AES therefore preceded XPS as a surface-sensitive spectroscopic technique. There are available fingerprint Auger spectra for the identification of surface species[24] with lineshapes enabling different states of an element to be recognised, as for example carbon present in gaseous CH_4, C_2H_4 and C_2H_2 and, when in the derivative form, in Mo_2C, SiC and graphite.

Although the two main anodes used in XPS are aluminium ($hv = 1486\,eV$) and magnesium ($hv = 1253\,eV$), synchrotron radiation sources have the advantage that they provide a quasi-continuous spectrum extending from the infrared into the X-ray region. The radiation source is strongly directional with linear polarisation and the ability to perform time-resolved experiments. Tunability is also an asset of synchrotron radiation for surface chemical information with EXAFS and SEXAFS benefiting. Thomas and his group at the Royal Institution[25] have obtained crucial information on the atomic environment of the active site in solid catalysts through X-ray absorption spectroscopy and X-ray diffraction using a tunable monochromatic beam of high-intensity X-rays from a synchrotron source.

Simultaneously with Siegbahn's development of XPS, Turner and Price were investigating how ultraviolet radiation could probe exclusively the valence orbitals (UPS) of gaseous molecules.[26] Slightly later, Bordass and Linnett applied UPS to the study of the adsorption of methanol at a tungsten surface under non-UHV conditions.[27] However, a prerequisite for exploring the application of XPS in surface science was coupling it with ultra-high vacuum (UHV) techniques and establishing quantitatively its surface sensitivity.

The first UHV-compatible spectrometer with multiphoton sources, UV and X-ray, became available in 1971 as ESCA-3 from Vacuum Generators and the establishment of their surface sensitivity for adsorbed species present at metal surfaces[28] over a wide temperature range,[29] including cryogenic temperatures (80 K). This opened up the possibility of defining the atomic nature of metal surfaces at the sub-monolayer level, the chemical state of adsorbates and in particular the interplay between molecular and dissociated states of adsorbates such as carbon monoxide and nitrogen, aspects relevant to the mechanisms of Fischer–Tropsch catalysis and ammonia synthesis.[30] There was also an insight into hitherto unexpected surface chemistry, the complexity of the chemisorption of nitric oxide at metal surfaces where N_2O is observed at cryogenic temperatures and the activation of adsorbates by chemisorbed oxygen being early examples.[31,32] As a surface-sensitive spectroscopic technique, XPS is unique in that it provides a complete surface analysis and, when used in a dynamic or real-time mode, with variable temperature facilities, it is ideally suited for studying the mechanisms of surface reactions. Its single drawback is that it does not bridge the pressure gap, although modifications to the ESCA-lab series of instruments can provide spectra in the presence of reactants at pressures of up to 2 mbar.[33] This was established in 1979.

One of the distinct advantages of XPS is that, through analysis of the core-level intensities, it can provide quantitative data on adsorbate concentrations. The following equation relates the surface concentration σ to the intensities of photoelectron peaks from sub-shells of a surface adatom Y_a and Y_s the integrated signal from a sub-shell of a substrate atom:

$$\sigma = \frac{Y_a \mu_s N \lambda \cos\phi d}{Y_s \mu_a M_s} \qquad (3)$$

where μ_a and μ_s are the sub-shell photoionisation cross-sections for the adatom and the substrate and are available from Schofield;[34] M_s is the molecular weight of the substrate, d is the density of the substrate, λ is the escape depth for the particular substrate sub-shell photoelectrons, ϕ is the angle of collection (with respect to the surface normal) of the photoelectrons and N is Avogadro's number. This equation is a modification of that suggested by Madey et al.[35] and its derivation and application are discussed elsewhere.[36] How to estimate overlayer thicknesses is also considered and attention is drawn to the need to separate out extrinsic and intrinsic contributions when plasmon loss features are present, as with aluminium core-level peaks.

A study of the reactivity of oxygen states at Ni(210) by XPS (Figure 2.2) where the latter was used in a "temperature-programmed desorption mode" provided an early stimulus for us to search for oxygen transients as the reactive states in oxidation catalysis. At Ni(210) the chemisorbed oxygen overlayer formed at 295 K was *inactive* in H abstraction from water at low temperatures, the water desorbing at 150 K, whereas the oxygen state formed at 77 K was active in hydroxylation.[37] The proposition was that the activity was associated with an O^- state, not fully coordinated within an "oxide" lattice and a precursor of the oxide O^{2-} state.

The availability of *in situ* XPS accompanying STM therefore provides both chemical characterisation of the adlayer and the concentrations of adatoms

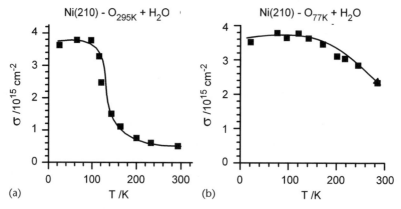

Figure 2.2 Reactivity of oxygen states chemisorbed at Ni(210) (a) at 295 K and (b) at 77 K to water adsorbed at 77 K. The "oxygen" concentration σ is calculated from the O(1s) spectra. The oxygen state preadsorbed at 295 K is unreactive with water desorption complete at 160 K whereas that at 77 K is reactive, resulting in surface hydroxylation.[37] (Reproduced from Refs. 37, 42).

present. It is surprising that very few studies have taken advantage of combining STM with *in situ* XPS.

2.7 The Dynamics of Adsorption

If we are to design the appropriate experimental strategy for providing meaningful data on the molecular events in kinetic aspects of surface reactions, the right questions need to be asked. Whether transient states are significant in the dynamics of oxygen chemisorption was an issue that we decided to address,[38] making use of the surface-sensitive spectroscopies available in the 1980s. We consider briefly the process of adsorption on solid surfaces, highlighting the individual events involved at the macroscopic level and emphasising the experimental prerequisites or limitations in developing a satisfactory model for a surface reaction.

When a molecule collides with a solid surface, a number of processes may occur. The molecule may be elastically scattered back into the gas phase or it may lose to the solid part of the translational component of its gas-phase kinetic energy normal to the surface and become trapped in a weakly adsorbed state. In very general terms, this would correspond to physical adsorption and the process of energy exchange referred to as surface accommodation. The molecule may not reach the ground state at its initial point of encounter with the surface but may diffuse (hop) to neighbouring sites, becoming de-excited as it moves. The molecules may also take up thermal energy from the lattice to overcome the activation energy for surface diffusion and in this sense is "hot", a concept we shall find to be commonplace in chemisorption studies. Eventually, the molecule will become chemisorbed. Very similar arguments can apply to molecular interaction which leads to dissociative chemisorption, the fragments of a diatomic molecule coming to rest in adjacent final states or well separated from each other. We consider next some quantitative implications of these concepts.

If N molecules strike 1 cm^2 of surface per second and have a residence time of t s, then the surface concentration σ is given by $\sigma = Nt$, where N is related to the gas pressure p by $N = p/2\pi mkT$, where m is the mass of the molecule, k the Boltzmann constant and T the temperature (K). At a pressure of 1 atm, the value of N for oxygen at 295 K is 2.7×10^{23} molecules cm^{-2} s^{-1}, which is approximately 10^8 times greater than the surface density of atoms at a solid surface ($\sim 10^{15}$ cm^{-2}). The implications for designing surface-sensitive experimental methods at well-defined atomically "clean" metal surfaces are clear: (a) UHV techniques are essential to minimize surface contamination from the gas ambient and (b) the experimental method should be capable of detecting at least 1% of the monolayer, *i.e.* 10^{13} cm^{-2}.

The surface residence time, t_{surf}, is related to the heat of adsorption, ΔH, and temperature, T, through a Frenkel-type relationship:

$$t_{\text{surf}} = t_0 \exp(\Delta H / RT) \qquad (4)$$

If we assume that $t_0 = 10^{-13}$ s (vibrational frequency)$^{-1}$, then for a heat of adsorption ΔH of 40 kJ mol^{-1} and a surface temperature of 295 K the residence time t_{surf} is 3×10^{-6} s and for 80 kJ mol^{-1} it is 10^2 s; as T decreases the value of the surface residence time increases rapidly for a given value of ΔH. Decreasing the temperature is one possible approach to simulating a "high-pressure" study in that surface coverage increases in both cases; the reaction, however, must not be kinetically controlled.

It is also important to distinguish between t_{surf}, the time a molecule is at the surface before desorbing, and the time it spends at a surface site, t_{site}, through surface diffusion or surface "hopping" (Figure 2.3). If the activation energy for diffusion is ΔE_{dif} then the time at a site t_{site} is given by eqn. (5), assuming that the pre-exponential factor is the same order of magnitude as that for desorption:

$$t_{site} = 10^{-13} \exp\left(\frac{\Delta E_{dif}}{RT}\right) \tag{5}$$

For a molecule characterised by a ΔH value of 40 kJ mol^{-1} and undergoing facile surface diffusion, i.e. a ΔE_{dif} value close to zero, then each molecule will visit, during its surface lifetime (10^{-6} s), approximately 10^7 surface sites. Since the surface concentration σ is given by $\sigma = N t_{surf}$, then for a ΔH value of 40 kJ mol^{-1} and $t_{surf} = 10^{-6}$ s at 295 K, the value of σ is $\sim 10^9$ molecules cm^{-2}. These model calculations are illustrative but it is obvious that no conventional spectroscopic method is available that could monitor molecules present at a concentration $\sim 10^{-6}$ monolayers. These molecules may, however, contribute, if highly reactive, to the mechanism of a heterogeneously catalysed reaction; we shall return to this important concept in discussing the role of transient states in catalytic reactions.

These are well-founded basic physico-chemical principles applied to molecules adsorbed at solid surfaces, but what is new is that they have been made relevant to understanding chemical reactivity by our experimental

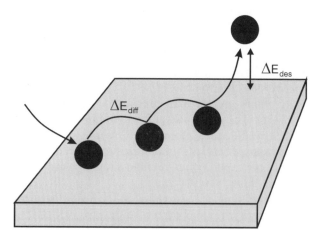

Figure 2.3 The energetics of a particle undergoing surface diffusion (ΔE_{dif}), desorption (ΔE_{des}) and the heat of adsorption (ΔH) $\simeq$ (ΔE_{des}).

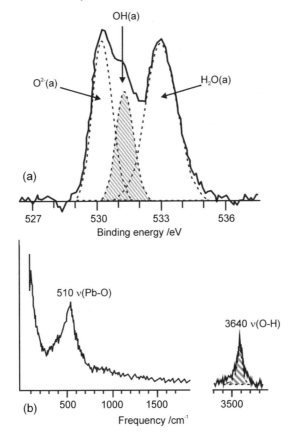

Figure 2.4 (a) O(1s) spectrum for the adsorption of a water–oxygen mixture at Pb(110) at 77 K and warming to 140 K with (b) electron energy loss spectrum confirming the presence of surface hydroxyls at 160 K when molecularly adsorbed water has desorbed. Both the "oxide" overlayer at Pb(110) and the atomically clean surface are unreactive to water. H abstraction was effected by transient $O^{\delta-}$ states, which were also active in NH_3 oxidation. (Reproduced from Refs. 40, 42).

coadsorption (mixture) studies. It is an approach that has not been made much use of in surface science, as emphasised by Sachtler.[39]

Figure 2.4 shows the advantage of combining XPS with HREELS for studying the coadsorption of water and oxygen at a Pb(110) surface at low temperatures.[40] The inherent inactivity of both Pb(110) and the "oxide" overlayer provided confidence to being able to attribute the catalytic oxidation activity to oxygen transient states $O^{\delta-}$, precursor states of the chemisorbed oxygen state O^{2-}(a).

Ammonia oxidation was a prototype system, but subsequently a number of other oxidation reactions were investigated by surface spectroscopies and high-resolution electron energy loss spectroscopy XPS and HREELS. In the case of ammonia oxidation at a Cu(110) surface, the reaction was studied under experimental conditions which simulated a catalytic reaction, albeit at low

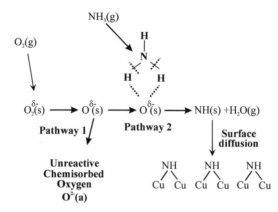

Figure 2.5 Dynamics of oxygen chemisorption at Cu(110) and Mg(0001) surfaces based on XPS and HREELS leading to oxidation of ammonia and imide chemisorbed species. The reactive oxygen is a transient state $O^{\delta-}(s)$. With ammonia-rich mixture Pathway 2 is dominant whereas for oxygen-rich mixtures Pathway 1 dominates at 295 K.

pressures. By varying the composition of the oxygen–ammonia *mixture* and "catalyst" temperature, reaction pathways could be controlled, emphasising how developing theoretical models for the reaction is intrinsically difficult if due consideration is not given to the dynamic nature of the reactive oxygen states. The following steps were proposed (Figure 2.5) for oxygen dissociation where the active oxygen was the $O^{\delta-}$ state in ammonia oxidation at Cu(110). It was envisaged as being analogous to a two dimensional gas reaction.[41,42]

$O_2(g) \rightarrow O_2(s)$	Thermal accommodation
$O_2(s) + e \rightarrow O_2^{\delta-}(s)$	1st stage of chemisorption; molecular transient
$O^{\delta-}_2(s) \rightarrow O^{\delta-}(a) + O^{\delta-}(s)$	Dissociative chemisorption with formation of "hot" transient $O^{\delta-}(s)$
$O^{\delta-}(s) \rightarrow O^{2-}(a)$	"Oxide" formation and loss of reactivity

The transient $O^{\delta-}(s)$ interacts with an ammonia molecule undergoing surface diffusion. A model was developed assuming that the following reaction occurs at an Mg(0001) surface:

$$NH_3(s) + O^{\delta-}(s) \rightarrow NH_2(a) + OH(a) \qquad (6)$$

In the gas phase, the reaction of O^- with NH_3 and hydrocarbons occurs with a collision frequency close to unity.[43] Steady-state conditions for both $NH_3(s)$ and $O^{\delta-}(s)$ were assumed and the transient electrophilic species $O^{\delta-}$ the oxidant, the oxide $O^{2-}(a)$ species poisoning the reaction.[44] The estimate of the surface lifetime of the $O^{\delta-}(s)$ species was $\sim 10^{-8}$ s under the reaction conditions of 298 K and low pressure ($\sim 10^{-6}$ Torr). The kinetic model used was subsequently examined more quantitatively by computer modelling the kinetics and solving the relevant differential equations describing the above

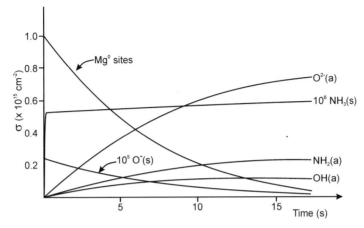

Figure 2.6 Variation of the concentration of surface species calculated from the differential equations describing the model for ammonia oxidation.[45] Efficient low-energy pathways to products are available through the participation of surface "transients" present at immeasurably low concentrations under reaction conditions. The NH$_3$ surface concentration is $\sim 10^{-6}$ ML. (Reproduced from Ref. 45).

mechanisms.[45] No assumptions were made concerning the steady-state conditions of the reacting species. The model established that with short lifetimes and assuming various activation energies for the surface diffusion (hopping) of ammonia (0, 7 and 14 kJ), the above reaction could be sustained (Figure 2.6) with reactants present at surface concentrations which could not be monitored by conventional spectroscopic methods but was kinetically "fast". It provided, at the very least, an impetus for the model to be scrutinised at the atom resolved level using STM, as it does not conform to neither the Langmuir–Hinshelwood or Eley–Rideal mechanisms for surface reactions.

However, for ammonia oxidation over Zn (0001), the kinetics indicated a precursor-mediated reaction, with rates increasing with decreasing temperature (Figure 2.7). A surface complex involving ammonia and dioxygen dissociated faster than dioxygen bond cleavage and the following mechanism was suggested;[46] it was an example of what was described as "precursor-assisted dissociation" by van Santen and Niemantsverdriet.[47]

$O_2(g) \rightarrow O^{\delta-}_2(s)$	Accommodation and first stage of chemisorption
$\frac{1}{2}O^{\delta-}_2(s) \rightarrow O^{\delta-}(s) \rightarrow O^{2-}(a)$	Inefficient reaction pathway to bond cleavage and chemisorption
$O^{\delta-}_2(s) + NH_3(s) \rightarrow [O^{\delta-}_2 \cdots NH_3]$	Complex formation
$[O^{\delta-}_2 \cdots NH_3] \rightarrow O^{2-}(a) + OH(a) + NH_2(a)$	Low-energy pathway to dioxygen bond cleavage

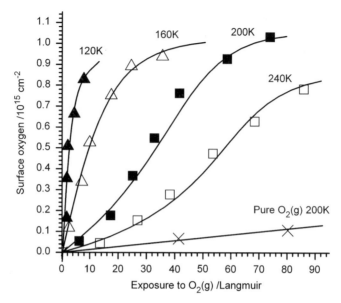

Figure 2.7 Surface oxidation at Zn(0001) in an ammonia-rich NH_3–O_2 mixture at 120, 160, 200 and 240 K compared with O_2(g) at 200 K as a function of O_2 exposure. An (NH_3–O_2) complex (transient) provides a low-energy pathway to dioxygen bond cleavage. The rate of dioxygen bond cleavage is increased by a factor of up to 300 in the presence of ammonia. (Reproduced from Ref. 46).

2.8 Summary

The experimental methods considered are those which provided experimental data that STM atom resolved evidence could profitably progress. The dynamics of oxygen chemisorption, including surface reconstruction and oxide formation, has been addressed as it underpins the mechanism of catalytic oxidation at metal surfaces. Spectroscopic studies of coadsorption making use of probe molecules to search for transient states have provided a model involving metastable oxygen states – both atomic $O^{\delta-}$(s) and molecular $O_2^{\delta-}$(s) – for the control of reaction pathways in catalytic oxidation. When dioxygen bond cleavage is slow, as for example at Zn(0001), then the dioxygen transient is the significant oxidant, providing via a precursor complex, a low-energy route to oxidation catalysis. By comparison, the chemisorbed "oxide" O^{2-}(a) overlayer is unreactive. What, then, can we learn from STM?

Table 2.1 gives some examples where spectroscopic studies (XPS and HREELS) provided evidence for the role of oxygen metastable transient states in oxidation catalysis.

Depending on the reaction conditions, in particular the ratios of oxygen to the reactant, there are two reaction pathways, one of which, oxygen-rich, leads to little or no activity at room temperature. The concepts implicit in the model

Table 2.1 Surface chemistry mediated via oxygen transients: evidence from surface spectroscopy.

Surface	Reactants	Observation
Mg(0001)	$O_2:NH_3$	Facile H abstraction
Mg(0001)	$O_2:C_3H_6$	C–H activation and H abstraction[48]
Al(pc)	$O_2:CO$	Low-energy pathway to C–O bond cleavage[49]
Zn(0001)	$O_2:C_5H_5N$	Facile route to dioxygen bond cleavage[50]
Cu(110)	$O_2:NH_3$	Selective oxydehydrogenation reactions giving N, NH or NH_2 species[51]
Cu(110)	$O_2:CH_3OH$	Selective for HCHO or surface formate[52]
Ni(210)	$O_2:H_2O$	Hydroxylation at low temperature[37]
Zn(0001)	$O_2:CH_3OH$	C–O bond cleavage at 80 K[53]
Zn(0001)	$O_2:H_2O$	Facile surface hydroxylation[54]
Ni(110)	$O_2:H_2O$	Surface hydroxylation at low temperatures
Ni(110) Cu(110)	$O_2:NH_3$	Oxydehydrogenation to give NH species[55]
Ag(111)	$O_2:H_2O$	Facile hydroxylation at low temperatures[56]

are non-traditional in the sense that they involve the participation of kinetically controlled metastable configurations of adsorbed oxygen, not previously considered part of traditional views on the dynamics of oxygen chemisorption at *metal* surfaces.

STM was seen as an approach that could provide definitive evidence regarding the formation, stability and surface lifetimes of oxygen transients, particularly through cryogenic studies, and whether analogies with radical-type two dimensional gas reactions were valid, a concept favoured by Semenov in 1951. He attributes to the catalyst surface the role of radical generator and chain terminator, the reaction steps leading to the products taking place between a radical and a "physically adsorbed", but otherwise unreactive, molecule. The implication of Semenov's view is the possibility that reaction need not involve two chemisorbed species; one can be a surface radical and mobile. This was discussed by H.A. Taylor in 1954 in the context of the Thon–Taylor Scheme[57] and also relevant to the model derived from coadsorption XPS studies in 1987.[38,44]

References

1. K. J. Laidler, in Catalysis, ed. P. H. Emmett, Reinhold, New York, 1954, 75; see also M. W. Roberts and C. S. McKee, *Chemistry of the Metal–Gas Interface*, Clarendon Press, Oxford, 1978.
2. P. Kisliuk, *J. Phys. Chem. Solids*, 1957, **3**, 95; 1958, 5, 78.
3. C. R. Arumainayagam, M. C. McMaster and R. J. Madix, *J. Phys. Chem.*, 1991, **95**, 2461.
4. D. A. King and M. G. Wells, *Surf. Sci.*, 1972, **29**, 454; *Proc. R. Soc. London, Ser. A*, 1974, **339**, 245.
5. G. Ehrlich, *J. Phys. Chem.*, 1955, **59**, 473J. *Phys. Chem. Solids*, 1956, **1**, 3.
6. C. M. Quinn and M. W. Roberts, *J. Chem. Phys.*, 1964, **40**, 237.
7. See references under Further Reading.

8. R. P. Eischens and W. A. Pliskin, *Adv. Catal.*, 1958, **10**, 1.
9. J. Pritchard and M. L. Sims, *Trans. Faraday Soc.*, 1970, **66**, 427; M. A. Chesters, M. Sims and J. Pritchard, *Chem. Commun.*, 1970, **1454**; R. G. Greenler and H. G. Tompkins, *Surf. Sci.*, 1971, **28**, 194.
10. N. Sheppard, in *Surface Chemistry and Catalysis*, ed. A. F. Carley, P. R. Davies, G. J. Hutchings and M. S. Spencer, Kluwer Academic/Plenum, London, 2002.
11. G. A. Somorjai and A. L. Marsh, *Philos. Trans. R. Soc. London, Ser. A*, 2005, **363**, 879, Discussion Meeting organised and edited by R. Mason, M. W. Roberts, J. M. Thomas and R. J. P. Williams.
12. Y. R. Shen, *Nature*, 1989, **337**, 519.
13. J. C. P. Mignolet, *Discuss. Faraday Soc.*, 1950, **8**, 326.
14. R. V. Culver and F. C. Tompkins, *Adv. Catal.*, 1959, **68**.
15. C. M. Quinn and M. W. Roberts, *Trans. Faraday Soc.*, 1964, **60**, 899; 1965, **61**, 1775; M. W. Roberts and B. R. Wells, *Discuss. Faraday Soc.*, 1966, **41**, 162.
16. A. U. MacRae, *Science*, 1963, **139**, 379; J. W. May, *Ind. Eng. Chem.*, 1965, **57**, 19.
17. J. B. Pendry, *Low Energy Electron Diffraction; the Theory and Its Application to the Determination of Surface Structure*, Academic Press, New York, 1974; S. Anderson and J. B. Pendry, *J. Phys. C*, 1980, **13**, 3547; see also references in Further Reading.
18. E. A. Wood, *J. Appl. Phys.*, 1964 **35**, No. 4, 1306.
19. C. S. McKee, L. V. Renny and M. W. Roberts, *Surf. Sci.*, 1978, **75**, 92; C. S. McKee, D. L. Perry and M. W. Roberts, *Surf. Sci.*, 1973, **39**, 176.
20. E. D. Williams, W. H. Weinberg and A. C. Sobrero, *J. Chem. Phys.*, 1982, **76**, 1150.
21. K. Heinz and L. Hammer, *J. Phys. Chem.*, 2004, **108**, 14579.
22. See for example D. P. Woodruff and T. A. Delchar, *Modern Techniques of Surface Science*, Cambridge University Press, Cambridge, 1986.
23. K. Siegbahn, *Philos. Trans. R. Soc. London, Ser. A*, 1986, **318**, 3, and references cited therein; Discussion Meeting at Royal Society, held in March 1985.
24. R. W. Joyner and M. W. Roberts, in *Surface and Defect Properties of Solids*, Eds: M. W. Roberts and J. M. Thomas, Chemical Society, London, 1975, Vol. **4**, p. 68.
25. J. M. Thomas, *Chem. Eur. J.*, 1997, **1**, 1557.
26. D. W. Turner, A. D. Baker, C. Baker and C. R. Brundle, *Molecular Photoelectron Spectroscopy*, Wiley, New York, 1970; A. W. Potts and W. C. Price, *Proc. R. Soc. London Series A*, 1971, **326**, 165.
27. W. T. Bordass and J. W. Linnett, *Nature*, 1969, **222**, 660.
28. C. R. Brundle and M. W. Roberts, in Discussion on the Physics and Chemistry of Surfaces, organised by J. W. Linnett, *Proc. R. Soc. London, Ser. A*, 1972, **331**, 383.
29. C. R. Brundle, D. Latham, M. W. Roberts and K. Yates, *J. Electron Spectrosc. Relat. d Phenom.*, 1974, **3**, 241.

30. K. Kishi and M. W. Roberts, *J. Chem. Soc. Faraday Trans. 1*, 1975, **71**, 1715; D. W. Johnson and M. W. Roberts, *Surf. Sci.*, 1979, **87**, L255.
31. D. W. Johnson, M. H. Matloob and M. W. Roberts, *J. Chem. Soc., Chem. Commun.*, 1978, **40**, *J. Chem. Soc., Faraday Trans. 1*, 1979, **75**, 2143.
32. C. T. Au, J. Breza and M. W. Roberts, *Chem. Phys. Lett.*, 1979, **66**, 340; M. W. Roberts and C. T. Au, *Chem. Phys. Lett.*, 1980, **74**, 472.
33. R. W. Joyner and M. W. Roberts, *Surf. Sci.*, 1979, **87**, 501.
34. J. H. Schofield, *J. Electron Spectrosc. Relat. Phenom.*, 1976, **8**, 129.
35. T. E. Madey, J. T. Yates and N. E. Erickson, *Chem. Phys. Lett.*, 1973, **19**, 487.
36. M. W. Roberts, *Adv. Catal.*, 1980, **29**, 55.
37. M. W. Roberts, A. F. Carley and S. Rassias, *Surf. Sci.*, 1983, **135**, 35.
38. C. T. Au and M. W. Roberts, *Nature*, 1986, **319**, 206; M. W. Roberts, *Chem. Soc. Rev.*, 1989, **18**, 451.
39. W. M. H. Sachtler, in *Surface Chemistry and Catalysis*, ed. A. F. Carley, P. R. Davies, G. J. Hutchings and M. S. Spencer, Kluwer Academic/Plenum, London, 2002, 207.
40. C. T. Au, A. F. Carley, A. Pashuski, S. Read, M. W. Roberts and A. Zeini-Isfahan, in *Adsorption on Ordered Surfaces on Ionic Solids and Thin Films*, ed. H.-J. Freund and E. Umbach, Springer, Berlin, 1993, 241.
41. A. Boronin, A. Pashusky and M. W. Roberts, *Catal. Lett.*, 1992, **16**, 345.
42. M. W. Roberts, *Surf. Sci.*, 1994, **299/300**, 769.
43. D. K. Bohme and F. C. Fesenfeld, *Can. J. Chem.*, 1969, **47**, 2718.
44. C. T. Au and M. W. Roberts, *J. Chem. Soc., Faraday Trans. 1*, 1987, **83**, 2047.
45. P. G. Blake and M. W. Roberts, *Catal. Lett.*, 1989, **3**, 399.
46. A. F. Carley, M. W. Roberts and Y. Song, *J. Chem. Soc., Chem. Commun.*, 1988, 267; *J. Chem. Soc., Faraday Trans.*, 1990, **86**, 2701.
47. R. A. van Santen and J. W. Niemantsverdriet, *Chemical Kinetics and Catalysis*, Plenum Press, New York, 1995.
48. C. T. Au, X.-C. Li, J.-A. Tang and M. W. Roberts, *J. Catal.*, 1987, **106**, 538.
49. A. F. Carley and M. W. Roberts, *J. Chem. Soc., Chem. Commun.*, 1987, 355.
50. A. F. Carley, M. W. Roberts and Y. Song, *Catal. Lett.*, 1988, **1**, 265.
51. B. Afsin, P. R. Davies, A. Pashusky, M. W. Roberts and D. Vincent, *Surf. Sci.*, 1993, **284**, 109.
52. P. R. Davies and G. G. Mariotti, *Catal. Lett.*, 1997, **43**, 261.
53. K. R. Harikumar and C. N. R. Rao, *Chem. Commun.*, 1999, 341.
54. M. W. Roberts, C. T. Au and A. R. Zhu, *Surf. Sci.*, 1982, **115**, L117.
55. G. K. Kulkarni, C. N. R. Rao and M. W. Roberts, *J. Phys. Chem.*, 1995, **99**, 3310.
56. A. F. Carley, P. R. Davies, M. W. Roberts and K. K. Thomas, *Surf. Sci. Lett.*, 1990, **238**, L467 *Appl. Surf. Sci.*, 1994, **81**, 265.
57. H. A. Taylor, *Ann. N.w Y. Acad. Sci.*, 1954, **58**, 198 N. N. Semenov, *Usp. Khim.*, 1951, **20**, 673.

Further Reading

K. Christmann, *Introduction to Surface Physical Chemistry*, Springer, New York, 1991.

G. Ertl and J. Küppers, *Low Energy Electrons and Surface Chemistry*, VCH, Weinheim, 1985.

D. P. Woodruff and T. A. Delchar, *Modern Techniques of Surface Science*, Cambridge University Press, Cambridge, 1986.

J. C. Rivière, *Surface Analytical Techniques*, Oxford Science Publications, Clarendon Press, Oxford, 1990.

R. Gomer, *Field Emission and Field Ionisation*, Oxford University Press, Oxford, 1961.

L. H. Germer and A. U. MacRae, A new low electron diffraction technique having possible applications to catalysis, *The Robert A. Welch Foundation Research Bulletin*, 1961, No. 11.

J. M. Thomas and W. J. Thomas, *Principles and Practice of Heterogeneous Catalysis*, VCH, Weinheim, 1997.

A.W. Czanderna and D. M. Hercules, *Ion Spectroscopies for Surface Analysis*, Plenum Press, New York, 1991.

G. A. Somorjai, *Principles of Surface Chemistry*, Prentice Hall, Englewood Cluffs, NJ, 1972.

M. A. Van Hove and S. Y. Tong, *Surface Crystallography by LEED*, Springer, New York, 1979.

R. J. H. Clark and R. E. Hester, *Spectroscopy for Surface Science*, Wiley, Chichester, 1998.

K. W. Kolasinski, Surface Science, *Foundations of Catalysis and Nanoscience*, Wiley, Chichester, 2002.

M. W. Roberts and C. S. McKee, *Chemistry of the Metal–Gas Interface*, Clarenden Press, Oxford, 1978.

C. S. McKee, M. W. Roberts and M. L. Williams, Defect structures studied by LEED, *Adv. Colloid Interface Sci.*, 1977, **8**, 29.

J. Pritchard, Reflection–absorption infrared spectroscopy, in *Chemical Physics of Solids and Their Surfaces*, Eds: M. W. Roberts and J. M. Thomas, Chemical Society, London, 1978, Vol. **7**, p. 157.

J. M. Thomas, E. L. Evans and J. O. Williams, Microscopic studies of enhanced reactivity at structural faults in solids, *Proc. R. Soc. London, Ser. A*, 1972, **331**, 417.

C. Defosse, M. Hovalla, A. Lycourghoitis and B. Delmon, Joint analytical electron microscopic and XPS study of oxide and sulfide catalysts, in *Perspectives in Catalysis*, ed. R. Larsson, C. W. K. Gleerup, 1981.

J. M. Thomas, The ineluctable need for *in situ* methods of characterising solid catalysts as a prerequisite to engineering active sites, *Chem. Eur. J.*, 1997, **3**, 1557.

M. M. Bhasin, Importance of surface science and fundamental studies in heterogeneous catalysis, *Catal. Lett.*, 1999, **59**, 1.

CHAPTER 3
Scanning Tunnelling Microscopy: Theory and Experiment

"Nothing tends so much to the advancement of knowledge as the application of a new instrument"

Sir Humphrey Davy

3.1 The Development of Ultramicroscopy

Feynman's prescient lecture on 29 December 1959 at the Annual Meeting of the American Physical Society at the California Institute of Technology declared "There's Plenty of Room at the Bottom". In it he challenged the scientific community to develop the technology to write and then read "*the entire 24 volumes of the Encyclopaedia Britannica on the head of a pin*", predicting the benefits of such miniaturisation to computing and also identifying electron microscopes as possible tools to achieve this feat. He acknowledged the fundamental problem of the resolution limit imposed by the wavelength of the probe involved; optical microscopy, for example, has a maximum resolution of ~ 500 nm and infrared microscopy of ~ 10 μm, but challenged the community to find a way around these problems. In fact, a way had already been suggested: E. H. Synge[1] discussed the possibility of using a 10 nm aperture to gain information on a molecular scale (easily surpassing Feynman's target) and even suggesting piezoelectric scanning as a means of controlling the microscope. Synge's ideas, however, were too far ahead of the day to be put into practice and they received little attention for almost 50 years.

The first demonstration of a scanning microscope with a resolution that defied the wavelength limit was by Ash and Nicholls[2] using microwaves and a sub-wavelength aperture in order to obtain what is called a near-field image, but again the concept failed to catch the general interest and the field remained quiescent until Binnig and Rohrer published an atomically resolved image of the Si(111) (7 × 7) reconstruction[3] (Figure 1.1). The experiment proved to be a watershed, partly because it resolved a longstanding issue on the nature of the surface reconstruction but also because of the sheer elegance of the image.

Binnig and Rohrer's advance arose from their common interest in thin oxide films and the localised inhomogeneities that affect the function of such films in devices. They recognised the lack of an appropriate probe with which such phenomena could be investigated and set about designing something suitable based on electron tunnelling, with which they both had some experience. Quantum tunnelling was postulated in the late 1920s by George Gamow, who used it to solve the problem of the mechanism of the decay of an atomic nucleus to give an alpha particle ("alpha decay"). In the early 1960s, it was observed that the tunnelling current measured between aluminium and lead films separated by an alumina film approximately 10 Å thick could give information on the electronic structure of the junction,[4,5] particularly with respect to superconductors. On the basis of these results, Young et al.[6] developed an instrument called a "Topografiner", which rastered a probe over a sample to give an image of the surface, the same principle on which the STM would eventually operate. Crucially, the resolution of the Topografiner was significantly worse than that of the electron microscopes then available.

Inspired partly by a 1976 paper on vacuum tunnelling by Thompson and Hanrahan,[7] Binnig and Rohrer devised the concept of a localised probe held within a few ångstroms of the surface by a feedback mechanism relying on the tunnelling of electrons between probe and surface; they realised within weeks that this proposal could provide not only spectroscopic information on the surface but potentially also topographical information. The patent for the first STM was submitted in 1979, but it was more than 2 years before an operational microscope was developed that was capable of demonstrating the anticipated characteristic exponential decay of the tunnelling current with probe sample separation and thereby establishing vacuum tunnelling. The problems faced by the team were those that still face today's practitioners of STM – sample preparation, tip preparation and the elimination of external noise. Some aspects of these problems will be discussed later in this chapter.

Within 3 months of achieving vacuum tunnelling, Binnig, Rohrer, Gerber and Weibel[8] recorded their first STM image showing monoatomic steps on CaIrSn$_4$ and Au(111) single-crystal surfaces (Figure 3.1). However, these preliminary images did not receive the attention they deserved and it was not until the following year[3] with the first real space images of the Si(111) (7 × 7) reconstruction (Figure 1.1) that the future potential of STM began to be appreciated by the surface science community.

Since the introduction of scanning tunnelling microscopy, a family of scanning probe microscopies (SPMs) have been developed (Table 3.1), with three main branches resulting from three different types of probe. All of the methods have in common the ability to image surfaces in real space at nanometre or better resolution, are straightforward to implement and are relatively low in cost.

Technologically, the most important member of the scanning probe family is perhaps the atomic force microscope (AFM), which has found applications in

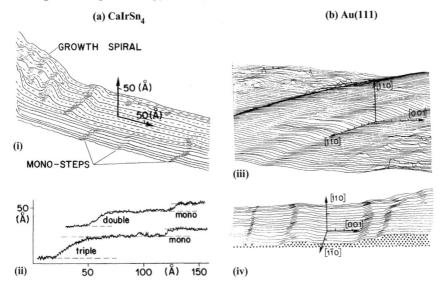

Figure 3.1 The first images recorded using a scanning tunnelling microscope to be published. Monoatomic steps are visible at (a) CaIrSn$_4$ and (b) Au(111) surfaces. (Reproduced from Ref. 43).

Table 3.1 Scanning probe microscopies and derivatives.

Scanning Tunnelling Microscopy (STM)

  Scanning tunnelling spectroscopy (STS)
  Spin-polarised scanning tunnelling microscopy (SP-STM)
  Magnetic force scanning tunnelling microscopy (MF-STM)

Atomic Force Microscopy (AFM)

  Non-contact atomic force microscopy (NC-AFM)
  Magnetic force microscopy (MFM)
  Electrical force microscopy (EFM)
  Lateral force microscopy (LFM)
  Friction force microscopy (FFM)

Scanning Near-field Optical Microscopy (SNOM)

fields as wide ranging as engineering, materials science and polymers and in particular in studying biological samples. AFM was developed by Binnig, Quate and Gerber[9] a few years after the advent of STM in an effort to resolve the latter's biggest drawback, its inability to image non-conducting surfaces. Unlike STM, AFM does not rely upon a tunnelling current, which requires a conducting sample; rather, the local probe is brought into proximity with the surface until it experiences an attractive (van der Waals) or repulsive (electron

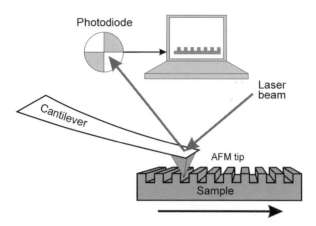

Figure 3.2 The essential elements of an atomic force microscope. The sample is moved beneath a tip mounted on a cantilever; a laser beam reflected off the back of the tip and on to a photodiode amplifies deflections of the cantilever.

repulsion) force. The probe is scanned over the surface in much the same way as in STM with the surface topography deduced from the extent to which the probe is deflected. The probe, a sharp tip often made from silicon nitride, is mounted on a cantilever and the light from a laser is reflected off the back of it on to a photodiode. Changes in the deflection of the probe are measured by the amplified motion of the light beam on the detector (Figure 3.2).

AFM can be used to study sample surfaces in air, liquid or vacuum and its ability to study non-conducting samples gives it very wide applicability. However, the original design involves simply dragging the tip across the surface; this *"contact"* or C-AFM has the major drawback of applying considerable lateral force to the samples during scanning, making it unsuitable for imaging "softer" samples such as biological specimens or polymers. The limitation of C-AFM can be overcome by increasing the tip–sample separation to between 5 and 15 nm so that the tip–sample interaction is governed principally by the attractive van der Waals forces. In this *"non-contact"* or NC-AFM mode, the cantilever oscillates at close to its resonant frequency and the topography of the surface is derived from changes in the amplitude, frequency and phase of the cantilever as the tip is scanned over the surface at a constant distance. The effect is to reduce the lateral forces on the sample to a minimum and has the further advantage of improved resolution over contact mode AFM; in fact, NC-AFM can achieve close to atomic resolution. The non-contact mode is only applicable to samples in solution or in vacuum since in air all samples have a condensed layer of liquid at their surface and under these conditions capillary forces dominate the interactions between sample and tip. Tapping AFM was developed to overcome this limitation.[10] In this mode, the cantilever also oscillates at close to its resonant frequency but approaches the sample much closer than in the non-contact mode; in effect, the tip taps the surface during scanning. Because the force exerted on the sample is always perpendicular to

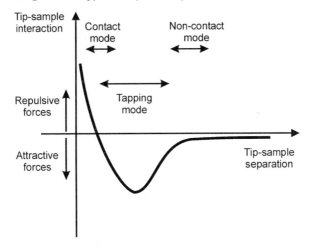

Figure 3.3 The interaction distances of the different modes of AFM operation.

the surface, the high lateral forces of contact mode AFM are avoided. The interaction distances for the different AFM modes are shown in Figure 3.3. AFM has since been adapted to measure a large variety of different forces between the surface and the tip, including the strength of magnetisation (MFM) and the frictional forces exerted by the surface on a scanning tip.

Scanning near-field optical microscopy (SNOM) is a separate branch of the SPM family and a direct descendent of the hypermicroscope proposed by Synge.[1] The idea behind this form of microscopy is to avoid the Abbé diffraction limit that restricts the resolving power of a normal imaging system to a function of the light wavelength. In practice, this means that it is not possible to resolve structures with dimensions smaller than half the wavelength of the light used. However, by keeping the probe–sample distance smaller than the size of the aperture, in other words by working in the "near field", the light from the source does not have the opportunity to diffract before interacting with the sample. In this case, the resolution of the microscope is determined by the diameter of the aperture. As in the case of the other probe microscopies described above, the aperture is scanned over the sample, with nanometre resolution. The image of the surface is obtained with a conventional "far-field" microscope detector and can include spectroscopic information. A nice practical demonstration of this effect was given by Danzebrink[11] and is reproduced in Figure 3.4. A SNOM tip is moved progressively closer to a surface, and as it approaches the near-field region the image resolves the surface features at a resolution of better than $\lambda/13$.

3.2 The Theory of STM

The basis of the scanning tunnelling microscope, illustrated schematically in Figure 3.5, lies in the ability of electronic wavefunctions to penetrate a potential barrier which classically would be forbidden. Instead of ending abruptly at a

Figure 3.4 Distance dependence of SNOM. A series of scanning transmission images are shown of a 20 nm high gold/palladium test structure on a silicon wafer using a source wavelength of 1064 nm. The scan height is sequentially reduced from ~300 nm (left image) to ~10 nm (right image). In the final image, the lateral resolution is ~80 nm, corresponding to $\lambda/13$, about one order of magnitude better than in conventional optical microscopy. (Reproduced from Ref. 11).

surface, the electron wavefunction decays exponentially and, because of this extension, electrons can jump into unoccupied states of a second conductor (the STM tip) if it approaches the surface sufficiently closely. An applied potential difference between the two conductors leads to a tunnelling current with a magnitude of nanoamperes. The tunnelling current, I_T, is directly related to the probability of electrons crossing the barrier and decays exponentially with the tip–sample separation z:

$$I_T \propto e^{-2\kappa z}$$

where

$$\kappa = \sqrt{\frac{2m(V_B - E)}{\hbar^2}}$$

E is the energy of the electrons, V_B is the vacuum energy, m is the mass of the electron and $\hbar$ is Planck's constant divided by 2π. $(V_B - E)$ is the local potential barrier height, which to a first approximation is the work function ϕ; for metal surfaces this is typically 4–5 eV.

If a voltage V is applied between the tip and sample, electrons within an energy eV (where e is the charge on an electron) of the Fermi level are able to tunnel through the barrier. Taking $V_B - E$ as 5 eV gives $\kappa \approx 1 \text{ Å}^{-1}$, thus the tunnelling current decreases by a factor of 10 for every ångstrom away from the surface. This also suggests that by changing the applied voltage, electrons from

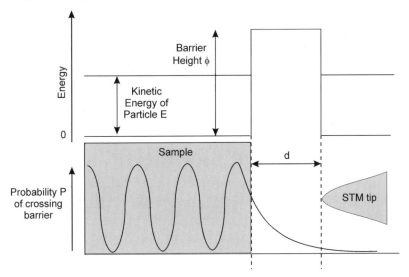

Figure 3.5 Graphical representation of the quantum mechanical tunnelling effect between tip and sample. The probability P of a particle with kinetic energy E tunnelling through a potential barrier ϕ is shown as a function of sample–tip separation z.

different energy levels can be probed – a form of tunnelling spectroscopy (STS); this is discussed below.

3.3 The Interpretation of STM Images

For most cases discussed in this book, STM images are interpreted from an assumed relationship between tunnelling current and surface topography and generally also on an understanding, from other experimental techniques, of the surface chemistry. However, in some cases, particularly at very high resolution, STM images can be ambiguous and a stronger theoretical understanding of the derivation of the image is required. The question of what is meant by the *topography* of a surface at the atomic level was raised in the early 1960s by researchers investigating tunnelling between superconducting metal plates separated by thin oxide films. Giaever[4] and Nicol et al.[5] found that they could quantitatively account for their data if they assumed that the tunnelling current was directly related to the density of electron states in the metal. This assumption was given theoretical support by Bardeen[12] shortly afterwards and his theory was applied to STM by Tersoff and Hamann.[13] A detailed description of the theory is given in a recent treatise by Gottlieb and Wesoloski.[14]

Theoretical models of STM images initially treated the STM tip as a point source of current, since, from the theoretician's point of view and despite all the efforts put in by experimentalists (see the discussion later), "little is known

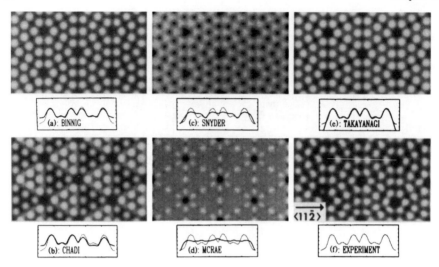

Figure 3.6 A comparison of an experimentally obtained STM image and line profile (f) with those calculated[15] from different Si(111) 7 × 7 models. In the line profiles underneath the image the dotted lines are the experimentally obtained data from (f) and the solid lines are the equivalent profiles from different structural models: (a) Binnig et al.;[3] (b) Chadi;[44] (c) Snyder;[45] (d) McRae and Petroff;[46] and (e) Takayanagi et al.[47] Very good agreement is obtained with Takayanagi et al.'s model. (Adapted from Tromp et al.[15]).

about the structure of the tunnelling probe tip, which is at present prepared in a relatively uncontrolled and non-reproducible manner."[13]

With a point source of tunnelling current and for low sample voltages, an STM image can be calculated from the density of states of the surface. An early example of an application of this approach is the work of Demuth and coworkers,[15] who calculated the STM images to be expected from competing models for the Si(111) 7 × 7 reconstruction and compared them with experiment. The results clearly favoured one model over all of the others and this has since become the generally accepted model[15] for the system (Figure 3.6).

Although point source models of an STM tip produce reasonably accurate qualitative models of STM images, quantitative results require more realistic tip models and these started to become possible in the early years of this decade. DFT calculations are used to construct the electronic structure of tungsten pyramids or a tungsten film. Frequently these are modified with a single atom of the substrate since these often give results closer to the experimental images. A detailed discussion of the theory of STM imaging is available in several reviews.[14,16,17]

3.4 Scanning Tunnelling Spectroscopy

A major deficiency in scanning tunnelling microscopy is the absence of direct chemical information, but the dependence of the tunnelling current on the local density of states near the Fermi level gives the possibility of probing the density of states by measuring the tunnelling current as a function of the applied bias – potentially a highly localised spectroscopy. The simplest application of this concept is to compare images of a surface with different tip–sample polarities; tunnelling to the sample from the tip probes empty states at the surface, whereas tunnelling from the surface to the tip probes filled states at the surface. STM images from silicon, for example, show dramatic changes with changed bias due to the filled dangling bonds present at the clean surface from which tunnelling is easily achieved but which hinder electron tunnelling to the sample. Figure 3.7 shows clearly that adsorbing an accepting molecule on top of the silicon enables the STM to tunnel in both directions but producing very different images under positive and negative sample bias.

Examples of STS that are related to catalysis include the work of Goodman and co-workers,[18–19] who have studied the electronic structure of palladium and gold nanoparticles on TiO_2 as a function of nanoparticle size using I–V curves

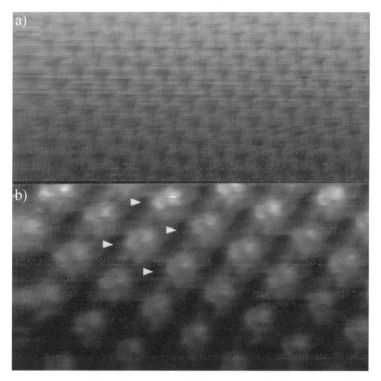

Figure 3.7 $11 \times 11\,nm^2$ STM image showing the sample bias-dependent imaging of cobalt phthalocyanine (CoPc) at an Ag/Si-$\sqrt{3}$ surface. Part (a) of the image was recorded with a bias voltage of -1.5 V and a tunnelling current of 0.5 nA and shows the Ag/Si-$\sqrt{3}$ surface. Part (b) of the image was recorded with a bias voltage of $+1.5$ V and a tunnelling current of 0.5 nA. This shows the CoPc molecular layer. The dark holes between molecules are arrowed. (Reproduced from Ref. 48).

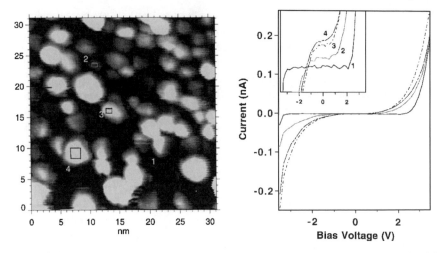

Figure 3.8 Scanning tunnelling micrograph of Pd(1.2 ML)/MgO(100)/Mo(100) with corresponding STS acquired at the indicated regions on the surface (boxes 1, 2, 3 and 4). (Reproduced from Ref. 18).

obtained at specific positions on nanometre-resolved images (Figure 3.8). Their results suggest the development of a bandgap in the metal particles as the mean diameter reduces from 10 to 4 nm. This is the same size domain in which gold particles are thought to become catalytically active for many processes. Stipe, Rezaei and Ho[20] developed the concept of atomically resolved spectroscopy even further using STM as a molecularly resolved version of inelastic electron tunnelling spectroscopy (IETS). The latter technique monitors the energy lost by tunnelling electrons to vibrational modes of molecules trapped at the interface of two electrodes. The vibrational signal is recorded from the second derivative of the tunnelling current with gap voltage, and to achieve this with STM required exceptionally high-quality data; the same area of surface was scanned for up to 10 h. Using this approach, Stipe *et al.* were able to distinguish between individual molecules of ethyne and deuterated ethyne (Figure 3.9). However, despite its obvious appeal, the demanding experimental conditions required for this experiment mean that it is unlikely to become a commonplace approach in the near future.

3.5 The STM Experiment

The basic features of the STM experiment are shown in Figure 3.10. A sharp tip is positioned using piezoelectric crystal drives within a few ångstroms of the surface. A potential difference is applied between sample and tip and an image obtained by rastering the tip across the surface. In the simplest implementation, the tip is controlled in one of two modes: "constant height", where the tip is held at a fixed distance from the surface and a direct image of the tunnelling current is produced, and "constant current", where a negative feedback circuit

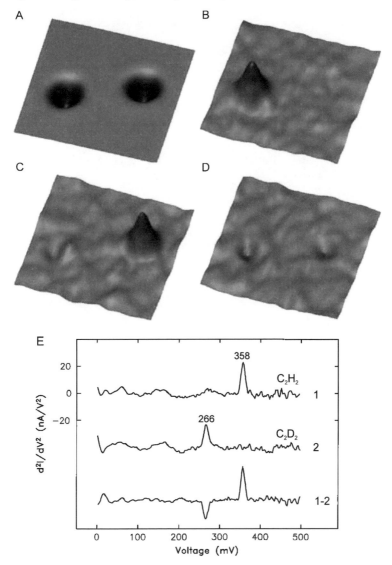

Figure 3.9 Spectroscopic spatial imaging of C_2H_2 and C_2D_2. (A) Regular (constant current) STM image of a C_2H_2 molecule (left) and a C_2D_2 molecule (right). Data are the average of the STM images recorded simultaneously with the vibrational images. The imaged area is 48×48 Å. d^2I/dV^2 images of the same area recorded at (B) 358, (C) 266 and (D) 311 mV. The symmetrical, round appearance of the images is attributable to the rotation of the molecule between two equivalent orientations during the experiment. (E) d^2I/dV^2 spectra for C_2H_2 (1) and C_2D_2 (2), taken with the same STM tip. (Reproduced from Ref. 20).

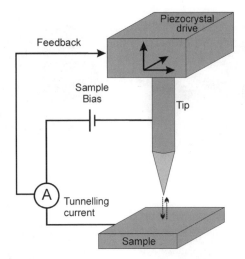

Figure 3.10 Typical STM experimental arrangement with negative feedback circuit to the piezoelectric crystal controlling movement in the z-direction.

retracts and extends the tip to maintain a constant current as it scans. In the latter mode, it is the retraction and extension of the piezo drive that are measured and although this mode generally results in a slower scanning speed it also reduces the risk of the tip crashing into surface steps. More recent implementations use a hybrid of the two modes where the degree of feedback is modified depending on the roughness of the surface. In general, at low magnifications, where larger areas are being scanned and multiple step edges are likely to be encountered, the constant current mode is preferred whereas when scanning a flat terrace at atomic resolution the constant height mode will give improved results. With metals and semiconductors, atomically resolved images were appearing regularly in the literature by the end of the 1980s. Oxides, however, presented a very different challenge and the application of atomically resolved techniques in this area has been much slower to develop.

The relative simplicity of the STM experiment is reflected by the fact that the microscope was commercially available within 5–6 years of its invention and by the end of the 1990s was a mature technology. Development has continued, however, with efforts being made to extend the range of conditions under which STM can be applied, in particular at high pressures and temperatures and also to improve the rate of image acquisition with the aim of enabling the STM to monitor surface structural changes in real time.[21] Results from several of these studies are discussed in later chapters.

3.6 The Scanner

The scanner is the heart of the tunnelling microscope, controlling the x, y and z motion of the tip relative to the sample. The requirements that the scanner must meet are severe: to obtain images of the surface with atomic resolution. The tip must be controlled to within an accuracy of less than 1 Å in the plane of the surface and better than 0.05 Å perpendicular to the surface. In addition, the scanner must have a high resonant frequency to reduce noise and to permit efficient feedback between signal and scanner. As long ago as the 1930s, Synge[22] proposed using piezoelectric materials to control the lateral positioning of his "hypermicroscope"; today, virtually all tunnelling microscopes rely on the commercially available PZT [Pb(Zr, Ti)O$_3$] piezoelectric ceramic to control the positioning of the tunnelling tip. The advantage of these materials is that in addition to displaying the required accuracy for atomic-scale scanning, they also have an almost linear dependence of lateral motion on voltage and negligible creep (at least at the low electric fields required in this application). Three types of piezoelectric actuators are used: bar, tube and stacked disc in a number of different scanner designs, one of the most popular being the tube scanner which has the advantage of compact size and simplicity. In this design, the piezoelectric crystal is formed into a hollow tube; the inner wall controls the z displacement whereas the x–y scanning motion is controlled by the outer walls. These are divided into four equal sections; an opposing voltage applied to two opposite areas on the tube results in a deformation of the tube perpendicular to the tube's axis. A number of researchers have studied the performance of piezoelectric tube scanners.[23–26]

3.6.1 Sample Approach

The initial stages of the STM experiment require the positioning of the tip in proximity of the surface such that a tunnelling current can be detected; this often means moving the tip by several micrometres or even millimetres. The piezoelectric materials used for scanning are not suitable for this initial approach and most instruments therefore contain a second "coarse positioning" driver; frequently this is also a piezoelectric material in a "stick–slip" kind of design.[27]

3.6.2 Adaptations of the Scanner for Specific Experiments

Amongst surface-sensitive techniques, STM is almost unique in being capable of studying systems under pressures ranging from ultra-high vacuum to several atmospheres and from liquid helium temperatures to over 1000 K. In recent years, a number of groups have developed specific STM instruments to study surfaces under these conditions. High pressures were the first to be tackled and the results of these experiments are discussed in Chapter 7. Generally, these systems have been designed with the STM scanner contained within the high-pressure chamber,[28–30] but one exception is the system designed by Frenken and co-workers,[31] in which a Viton seal protects the STM scanner from the high-pressure gases and only the tip protrudes through the seal (Figure 3.11). The system is capable of imaging surfaces under a reactive flowing gas mixture

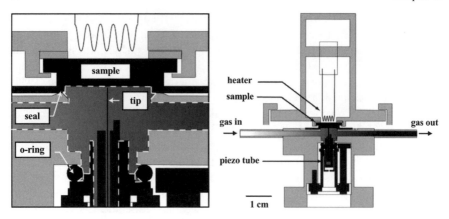

Figure 3.11 Schematic diagram of a high-pressure/high-temperature STM design in which only the tip is exposed to reactive gases. The instrument can image a surface, while it is active as a catalyst, under gas flow conditions at pressures up to 5 bar and temperatures up to 500 K. The volume of the cell is 0.5 ml. (Reproduced from Ref. 31).

and, because the scanner is also thermally insulated from the sample environment, at much greater temperatures than would normally be possible with a piezoelectric material-based instrument.

3.7 Making STM Tips

The importance of the STM tip was recognised from the earliest STM experiments; in their first paper detailing a scanning tunnelling microscope,[8] Binnig *et al.* estimated the radius r of a spherical tip necessary to resolve a monoatomic surface step as approximately $3r^{1/2}$. This implies that to image features at the nanometre level, a tip with a radius of close to 1 nm is required and ideally the tip should terminate in a single atom. However, this requirement does not turn out to be as stringent as first supposed; Binnig and Rohrer ground their first STM tips mechanically and it has since been shown that simply cutting a thin wire at an angle with a pair of scissors will create tips capable of imaging nanometre-scale objects.[32] The explanation for this unexpected resolution is thought to be that a series of "minitips" extend from the overall tip surface and, as a result of the exponential dependence of tunnelling current on distance, the longest of these minitips dominates the tunnelling (Figure 3.12). However, it was soon recognised that although mechanically ground or cut tips could be functional, they are generally short-lived and prone to producing multiple images arising from multiple minitips interacting with the surface simultaneously. STM practitioners adopted methods of producing sharp tips which had been developed originally for field electron emission microscopy (FEEM) and field ion microscopy (FIM). The former requires tip radii of the order of a 1 μm whereas the latter need to be even sharper, with radii typically of the order of 100 nm.

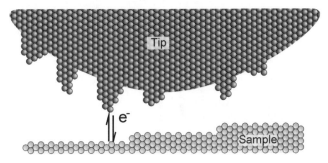

Figure 3.12 Illustration of the supposed structure of a typical STM tip at the atomic level showing a number of asperities through which tunnelling might be expected to occur.

Three criteria are generally recognised as being necessary for a "good" STM tip: (i) a single tip (split or multiple tips giving rise to overlapping images); (ii) a low aspect ratio (length/width ratio) away from the tip in order to give good mechanical stability and thereby reduce oscillations during scanning; and (iii) a high aspect ratio near the end of the tip, which improves access of the tip to rough areas of a surface. A large number of tip preparations have been described in the literature, including electrochemical etching, ion milling, mechanical grinding and cutting. A critical review of some of these methods has been given by Melmed.[33]

By far the most common method of STM tip preparation is electrochemical etching, which, in its simplest application, involves suspending a 0.5–1 mm diameter wire in a low-concentration electrolyte (0.3 3 M) such as NaOH or KOH and applying 1–10 V AC or DC between it and a counter electrode (Figure 3.13). Overall the etching reaction corresponds to

$$W + 2H_2O + NaOH \rightarrow 3H_2 + Na_2WO_4$$

The current is maintained until the end of the wire in the solution has been etched away or drops off. Stopping the etch at this point is critically important since any further etching acts to blunt the new tip and elaborate electronic methods have been applied to achieve a quick cut-off of current. An alternative approach, the "lamellae" method, involves suspending the electrolyte in a ring-shaped counter electrode with the wire to be etched hanging through it. Gravity separates the two pieces of wire at the end of the etching process automatically. The one drawback of this approach is that the required tip drops into a collection beaker and could be damaged. Tethering the tip to a soft platinum spring[34] prevents it from dropping but has the disadvantage of introducing lateral and vertical forces on the wire as it is etched, thereby distorting the final tip.

A further improvement was introduced by Weiss and co-workers,[35] who connected the circuit between the lamellae and the lower portion of the wire through an electrolyte in a conducting beaker. This avoids introducing any forces on the tungsten wire whilst retaining the automatic etching current

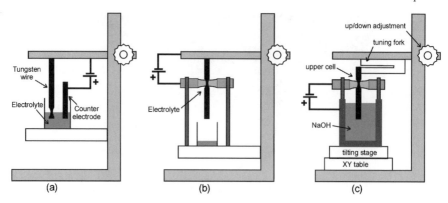

Figure 3.13 Three methods of chemically etching metal tips for STM. In (a) the current cut-off is manually or electronically triggered when the end of the etched wire falls; the finite time delay inherent in this approach results in a blunting of the final tip as etching continues after separation. (b) This shows an adaptation in which the etching current is automatically cut off when the lower portion of the wire drops – it is the lower portion that is used as an STM tip. (c) This shows an improved design in which the etching current is fed to the lower portion of the tungsten wire through an electrolyte held in a conductive beaker. In this case the upper portion of the etched wire is kept.

cut-off and, since it is the top portion of the wire that is retained, it also avoids the possibility of damage to the tip at separation. An electron micrograph showing a tip etched using this method is given in Figure 3.14. There are examples of even more sophisticated and specialised tip preparation methods available in the literature, for example, field ion microscopy can be used to "machine" tips to an atomically precise geometry.[36] These rather demanding methods are generally used only in very specific applications.

3.7.1 Tip Materials

Tungsten is the most widely used metal for STM tips; it is mechanically hard, easily available and can be etched quickly and easily. However, tungsten does have drawbacks; in particular, during electrochemical etching an oxide layer up to 10 nm thick develops at the very end of the tip.[37] This layer can contain a variety of other elements including potassium and carbon and is thought to have a detrimental effect on image quality. Various methods have been proposed to remove the oxide layer from tungsten tips, including etching in HF, annealing at high temperatures and ion milling. The success of these different methods has generally been gauged from the tunnelling performance, but when Ottaviano *et al.*[38] used scanning Auger microscopy and SEM to compare the different tip cleaning approaches they all proved to be disappointing, the effect of high temperature annealing of tungsten tips in order to remove the tungsten oxide by sublimation being particularly surprising (Figure 3.15).

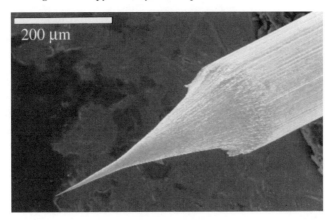

Figure 3.14 Scanning electron microscope image of a tip etched by the lamellae drop-off etching technique. (Reproduced from Ref. 35).

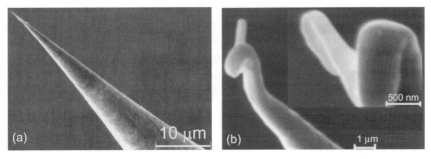

Figure 3.15 SEM images showing the effect of annealing tungsten STM tips using high sample bias (110 V) and tunnelling currents (50 nA). (a) Etched W tip before annealing; (b) tip after annealing. (Reproduced from Ref. 38).

Pt and Pt–Ir tips are less susceptible to contamination than tungsten and as a result are often used for STM systems operating at atmospheric or higher pressure.[28] Because these metals are softer than platinum, a simple cut is frequently used to make STM tips; however, the SEM images of Shapter and co-workers[32] show the significant differences between these cut tips and those that are subsequently etched in a mixed water–acetone solution (20:1) with $CaCl_2 \cdot H_2O$ as an electrolyte.

Gold as the most noble of metals is a very attractive metal for STM tip material, particularly for high-pressure systems,[28] but Kolmakov and Goodman[39] have reported that although gold tips are inert, their mechanical and thermal stability is not sufficient for imaging reactions at metal surfaces at high pressure. Similarly, Pt–Ir tips are relatively inert but are not stiff enough to withstand occasional contact with a surface. They recommend tungsten tips in a reducing atmosphere and polycrystalline tungsten tips coated with a thin oxide layer as a stiffness–reactivity compromise in oxidising atmospheres when atomic resolution is not required. In contrast, Wintterlin's group has more recently reported[40] stable tunnelling conditions in an oxygen atmosphere with tungsten as a tip material. Wintterlin et al.'s fast STM system also continues to use tungsten tips.[21]

Other materials used for STM tips have generally been prepared for specific applications, for example spin polarised STM, a magnetically sensitive imaging technique which requires tips constructed from ferromagnetic or antiferromagnetic materials.[41] Nickel, iron, chromium, chromium oxide-coated silicon, tungsten coated with iron and manganese–nickel or manganese–platinum alloys[42] have all been prepared for this type of application.

References

1. E. H. Synge, *Philos. Mag.*, 1928, **6**, 356.
2. E. A. Ash and G. Nicholls, *Nature*, 1972, **237**, 510.
3. G. Binnig, H. Rohrer, C. Gerber and E. Weibel, *Phys. Rev. Lett.*, 1983, **50**, 120.
4. I. Giaever, *Phys. Rev. Lett.*, 1960, **5**, 464.
5. J. Nicol, S. Shapiro and P. H. Smith, *Phys. Rev. Lett.*, 1960, **5**, 461.
6. R. Young, J. Ward and F. Scire, *Rev. Sci. Instrum.*, 1972, **43**, 999.
7. W. A. Thompson and S. F. Hanrahan, *Rev. Sci. Instrum.*, 1976, **47**, 1303.
8. G. Binnig, H. Rohrer, C. Gerber and E. Weibel, *Phys. Rev. Lett.*, 1982, **49**, 57.
9. G. Binnig, C. F. Quate and C. Gerber, *Phys. Rev. Lett.*, 1986, **56**, 930.
10. Q. Zhong, D. Inniss, K. Kjoller and V. B. Elings, *Surf. Sci.*, 1993, **290**, L688.
11. H. U. Danzebrink, Distance dependence of near-field optical resolution, http://www.nahfeldmikroskopie.de/Spektroskopie/spektroskopie.html; accessed 11 October 2006.
12. J. Bardeen, *Phys. Rev. Lett.*, 1961, **6**, 57.
13. J. Tersoff and D. R. Hamann, *Phys. Rev. Lett.*, 1983, **50**, 1998.
14. A. D. Gottlieb and L. Wesoloski, *Nanotech.*, 2006, **17**, R57.
15. R. M. Tromp, R. J. Hamers and J. E. Demuth, *Phys. Rev. B*, 1986, **34**, 1388.
16. W. A. Hofer, A. S. Foster and A. L. Shluger, *Rev. Mod. Phys.*, 2003, **75**, 1287.
17. D. Drakova, *Rep. Prog. Phys.*, 2001, **64**, 205.
18. D. R. Rainer and D. W. Goodman, *J. Mol. Catal. A*, 1998, **131**, 259.
19. D. W. Goodman, D. C. Meier and X. Lai, in: *Surface Chemistry and Catalysis*, ed. P. R. Davies, A. F. Carley, G. J. Hutchings and M. S. Spencer, Kluwer, New York, 2002, p. 148.
20. B. C. Stipe, M. A. Rezaei and W. Ho, *Science*, 1998, **280**, 1732.
21. J. Wintterlin, J. Trost, S. Renisch, R. Schuster, T. Zambelli and G. Ertl, *Surf. Sci.*, 1997, **394**, 159.
22. E. H. Synge, *Philos. Mag.*, 1931, **11**, 65.

23. S. Y. Yang and W. H. Huang, *Rev. Sci. Instrum.*, 1998, **69**, 226.
24. J. Tapson and J. R. Greene, *Rev. Sci. Instrum.*, 1997, **68**, 2797.
25. M. E. Taylor, *Rev. Sci. Instrum.*, 1993, **64**, 154.
26. R. G. Carr, *J. Microsc.*, 1988, **152**, 379.
27. S.-I. Park and R. C. Barrett, in: *Scanning Tunneling Microscopy*, ed. J. A. Stroscio and W. J. Kaiser, Academic Press, San Diego, 1993, p. 31.
28. G. A. Somorjai, *Appl. Surf. Sci.*, 1997, **121**, 1.
29. J. A. Jensen, K. B. Rider, Y. Chen, M. Salmeron and G. A. Somorjai, *J. Vac. Sci. Technol. B*, 1999, **17**, 1080.
30. E. Laegsgaard, L. Osterlund, P. Thostrup, P. B. Rasmussen, I. Stensgaard and F. Besenbacher, *Rev. Sci. Instrum.*, 2001, **72**, 3537.
31. B. L. M. Hendriksen, S. C. Bobaru and J. W. M. Frenken, *Top. Catal.*, 2005, **36**, 43.
32. B. L. Rogers, J. G. Shapter, W. M. Skinner and K. Gascoigne, *Rev. Sci. Instrum.*, 2000, **71**, 1702.
33. A. J. Melmed, *J. Vac. Sci. Technol. B*, 1991, **9**, 601.
34. F. W. Niemeck and D. Ruppin, *Z. Angew. Phys.*, 1954, **6**, 1.
35. M. Kulawik, M. Nowicki, G. Thielsch, L. Cramer, H. P. Rust, H. J. Freund, T. P. Pearl and P. S. Weiss, *Rev. Sci. Instrum.*, 2003, **74**, 1027.
36. A. S. Lucier, H. Mortensen, Y. Sun and P. Grutter, *Phys. Rev. B*, 2005, **72**.
37. A. Cricenti, E. Paparazzo, M. A. Scarselli, L. Moretto and S. Selci, *Rev. Sci. Instrum.*, 1994, **65**, 1558.
38. L. Ottaviano, L. Lozzi and S. Santucci, *Rev. Sci. Instrum.*, 2003, **74**, 3368.
39. A. Kolmakov and D. W. Goodman, *Rev. Sci. Instrum.*, 2003, **74**, 2444.
40. M. Rossler, P. Geng and J. Wintterlin, *Rev. Sci. Instrum.*, 2005, **76**.
41. M. Bode, *Rep. Prog. Phys.*, 2003, **66**, 523.
42. S. F. Ceballos, G. Mariotto, S. Murphy and I. V. Shvets, *Surf. Sci.*, 2003, **523**, 131.
43. G. Binnig and H. Rohrer, *Surf. Sci.*, 1983, **126**, 236.
44. D. J. Chadi, *Phys. Rev. B*, 1984, **30**, 4470.
45. L. C. Snyder, *Surf. Sci.*, 1984, **140**, 101.
46. E. G. McRae and P. M. Petroff, *Surf. Sci.*, 1984, **147**, 385.
47. K. Takayanagi, Y. Tanishiro, M. Takahashi and S. Takahashi, *J. Vac. Sci. Technol. A*, 1985, **3**, 1502.
48. M. D. Upward, P. H. Beton and P. Moriarty, *Surf. Sci.*, 1999, **441**, 21.

CHAPTER 4
Dynamics of Surface Reactions and Oxygen Chemisorption

"Qualitative logic is a prerequisite of quantitative theory"

Anon

4.1 Introduction

With the availability of surface-sensitive spectroscopies, it became possible to examine some aspects of the Langmuir model, with oxygen reactivity and dynamics playing a significant role in the development of new concepts in surface kinetics. There were two pointers that led us to question at a Faraday Symposium held in Bath in 1986 whether kinetic models for surface reactions at metal surfaces could be described adequately in terms of the classical Eley–Rideal or Langmuir–Hinshelwood mechanisms.[1] First, and contrary to what was generally accepted, surface oxygen could act as a promoter, facilitating bond breaking at cryogenic temperatures,[2] and second, metastable or transient oxygen states could participate in and control reaction pathways in oxidation catalysis. In particular, O^- transients generated during the dynamics of oxygen dissociation were the reactive sites, they were not "fully chemisorbed", non-thermalised and with significant surface lifetimes; they were given the term "hot". What was also pertinent was that the formation of O^- in the gas phase is highly exothermic, $O(g) + e \rightarrow O^-(g)$, $\Delta H = -140 \text{ kJ mol}^{-1}$, whereas the formation of O^{2-} is highly endothermic and only stable in the "final oxide" O^{2-} state due to the contribution from the Madelung energy associated with surface reconstruction and the "surface oxide". It had also been recognised that the transition from O^- to O^{2-} states led to the "shut down" of catalytic oxidation activity. Reactive oxygen was also evident in XPS studies of nitric oxide coadsorbed with water at a Zn(0001) surface at 180 K; in this case the reactive oxygen was generated *in situ* through cleavage of the nitrogen–oxygen bond. Surface hydroxylation was complete, analysis of the O(1s) intensity indicating a concentration of 4.7×10^{14} OH species cm^{-2}. Thermalised preadsorbed chemisorbed oxygen at Zn (0001) is inactive.[3]

Support for the special reactivity of "hot" oxygen adatoms also came from Matsushima's temperature-programmed desorption study[4] of CO oxidation at Pt(111), when CO and O_2 were coadsorbed at low temperature with CO_2 desorbed at 150 K, the temperature at which O_2 dissociates. This temperature is some 150 K lower than that for CO_2 formation when oxygen is preadsorbed (thermally accommodated) at the Pt(111) surface.

Although chemical reaction dynamics had been under intense scrutiny in the 1980s by molecular beam studies, Mullins *et al.*[5] at IBM emphasised in 1991 that "*there had been very few studies of the dynamics of reactions involving more than one reactant*". In their studies of CO oxidation, Auerbach's group established under molecular beam conditions that if the oxygen atoms were supplied as an atomic beam, CO_2 formation at Pt(111) was highly efficient with oxygen that had not been thermally accommodated, which they suggested were in the ground $O(^3P)$ state.

An Editorial in *Cattech* in 1997 also highlighted the possible significance of "hot" atoms and transient reaction intermediates in catalysis, drawing attention to a very different concept that was emerging for interpreting chemisorption and surface reactivity, with radical-type reactions participating in the mechanism.[6] How, then, was this to be viewed at the atom resolved level through the availability of STM and what were the implications for the development of theoretical models? It was to address these questions that led us to acquire in 1997, through EPSRC funding, a specially designed STM from Omichron with *in situ* XPS for chemical information and cryogenic facilities. We gave priority to determining initially whether models derived from surface spectroscopy could be sustained at the atom resolved level.

The outstanding feature of the early STM studies by the Bessenbacker and Ertl groups of oxygen chemisorption at metal surfaces was the facile mobility of both substrate and oxygen adatoms in surface reconstruction. There were, however, no low-temperature studies and attention was given to the analysis of high oxygen coverages, "the oxide monolayer" and the development of structural models. Ertl's group, in a study of oxygen chemisorption at Al(111) at 300 K, described in a series of papers[7] the formation of "ordered patches" (islands) as a consequence of "hot" oxygen atoms, formed by dioxygen bond cleavage, undergoing rapid surface diffusion (Figure 4.1).

There was therefore a clear need to assess the assumptions inherent in the classical kinetic approach for determining surface-catalysed reaction mechanisms where no account is taken of the individual behaviour of adsorbed reactants, substrate atoms, intermediates and their respective surface mobilities, all of which can contribute to the rate at which reactants reach active sites. The more usual classical approach is to assume thermodynamic equilibrium and that surface diffusion of reactants is fast and not rate determining.

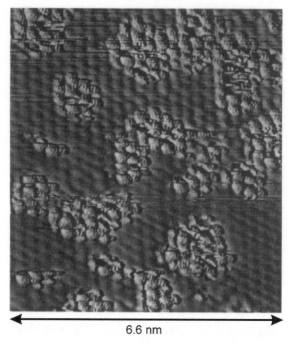

← 6.6 nm →

Figure 4.1 STM images of oxygen chemisorption at Al(111) at room temperature indicating the nucleation of "oxygen patches" after an exposure of 72 L. (Reproduced from Ref. 7).

4.2 Surface Reconstruction and "Oxide" Formation

Although thermodynamic data[8] predict that oxygen interaction with most metals should result in the formation of the thermodynamically stable "oxide overlayer" at room temperature, it is interesting to recall that work function and photoemission studies indicated that the "oxide overlayer" was likely to be defective and metastable. At a Faraday Discussion meeting in 1966, the work function and photoemission data for oxygen chemisorption at nickel were interpreted as involving a chemisorbed state Ni–O$^{\delta-}$ and two defective states Ni$_m$O and Ni$_x$O, the predominance of one or other of these states being controlled by oxygen pressure and temperature with the surface mobility of nickel playing a part[9] (see also Figures 2.1 and 2.2). We consider how concepts implicit in this model for oxygen chemisorption at metals, and in particular at Cu(110) and Ni(110), stand up to scrutiny by STM.

The first STM evidence for the facile transport of metal atoms during chemisorption was for oxygen chemisorption at a Cu(110) surface at room temperature;[10] the conventional Langmuir model is that the surface substrate atoms are immobile. The reconstruction involved the removal of copper atoms from steps [eqn (1)], resulting in an "added row" structure and the development of a (2 × 1)O overlayer [eqn (2)]. The steps present at the Cu(110) surface are

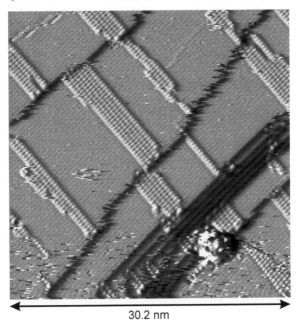

← 30.2 nm →

Figure 4.2 Development of (2 × 1)O strings at Cu(110) at 295 K with the strings bridging step-edges and emphasising that the growth mechanism involves the ends of the strings. There is no evidence for isolated oxygen adatoms at this coverage ($\theta = 0.25$), indicating their mobility. One isolated single string can be seen, pinned by two surface steps.

well defined and monatomic, being 0.128 nm high, whereas the step edges are often of irregular shape and "fuzzy", a consequence of the dynamic character of the surface. Due to the low activation energy for diffusion, the copper atoms are mobile at room temperature but are, however, trapped by chemisorbed oxygen atoms with the formation initially of isolated "strings" (Figure 4.2) which coalesce to form ordered (2 × 1) structures; the distance between the strings within the ordered structure is twice the Cu–Cu distance in the [110] direction (Figure 4.3). The anisotropic shape of the (2 × 1)O structures is a consequence of the strong Cu–O bonds within the strings compared with the interaction energy between the strings so that the probability of "trapping" a diffusing adatom at the end of a row is greater than at the side of a row, with copper removal from the step being rate-determining in string formation.

$$\text{Cu (step)} \rightarrow \text{Cu (terrace)} \tag{1}$$

$$\text{Cu (terrace)} + O^{\delta-}(s) \rightarrow -\text{Cu}-\text{O}-\text{Cu} \tag{2}$$

Step movement during chemisorption appears to be a general phenomenon. Real-time images observed (Figure 4.4) for chlorine chemisorption at Cu(110) indicate that nucleation takes place at a defect site, resulting in a single "string

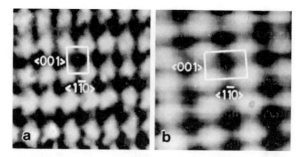

Figure 4.3 Atomically resolved STM image (1.5 × 1.5 nm) of a clean Cu(110) surface (a) before and (b) after the formation of a fully developed (2 × 1) oxygen adlayer at room temperature. (Reproduced from Ref. 10).

formation"; these strings (domains) develop and are 18 Å apart, reflecting a buckled surface which with time relax to give a c(2 × 2) structure.[11] Further exposure to HCl(g) results in further buckling of the surface (see Chapter 8).

The oxygen-induced reconstruction, observed with Ni(110), is also characterised at low oxygen coverage by the formation of –Ni–O–Ni–O– strings running along the [001] direction, *i.e.* perpendicular to the close-packed direction.[12] Ni–Ni bonds are broken, which is compensated for by a gain in chemisorption energy associated with the reconstructed "oxide" surface compared with the "clean" unreconstructed surface. Coexisting with these strings are (3 × 1) and (2 × 1) reconstructions (Figure 4.5), which develop locally and correspond to *local* oxygen coverages of 0.33 and 0.5 monolayers, respectively. At higher oxygen exposures the (2 × 1) added row structure is completed at the expense of the string structure, with another (3 × 1) added row structure developing corresponding to an oxygen coverage of 0.66 monolayers. Further structures are observed with increasing exposure corresponding to a (9 × 5) oxide and finally an epitaxial oxide overlayer of NiO(100). Besenbacher's group is of the view that the (9 × 5) structure is a two-layer structure with the Ni–Ni distance close to the Ni–Ni distance in NiO(100), the nickel atoms having rearranged compared with their positions in the metallic lattice. The smaller (2 × 1) strings are seen frequently to be mobile but become stabilised and immobile, when they grow into longer islands. It also follows that at the terraces there is in effect a two-dimensional gas involving mobile oxygen and metal atoms at both Cu(110) and Ni(110) surfaces. We shall see that this has a significant implication for the mechanism of surface reactions, including catalytic oxidation at metal surfaces. It was, however, an aspect that was highlighted by surface spectroscopic studies initiated to search for the existence of oxygen transient states present in the dynamics of oxygen chemisorption at metal surfaces,[13] using ammonia as a probe molecule.

Surface reconstruction, which had dominated much of surface science through LEED studies, was very much a central theme of STM in the early 1990s but with surprisingly little attention given to chemical reactivity and the origin of active sites in heterogeneous catalysis. This was in part due to the lack of *in situ* chemical information that could be directly related to the STM images and

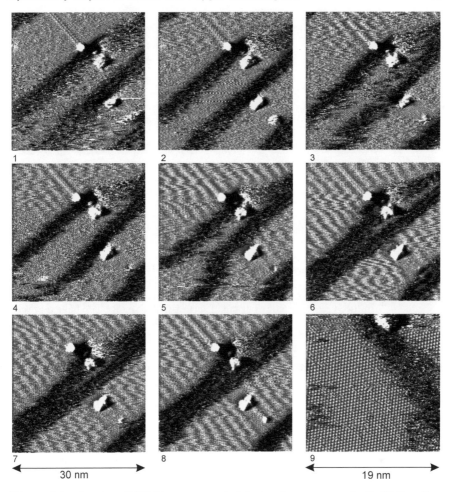

Figure 4.4 A series of STM images recorded during the exposure of a Cu(110) surface to hydrogen chloride at 295 K resulting in the formation of domains accompanied by step movement (1–8). With time this surface at 295 K relaxes to give a well-ordered c(2 × 2)Cl overlayer (9).

also that these images referred almost exclusively to just room temperature – a serious disadvantage to unravelling the mechanism of a dynamic process. These adsorbate-induced reconstructions and step movement are a common phenomenon, not confined to oxygen as an adsorbate; other examples are carbon and sulfur at Ni(100) and caesium at Cu(110), all at room temperature.[14]

4.3 Oxygen States at Metal Surfaces

Oxygen chemisorption at cryogenic temperatures provided the clue for the presence of metastable reactive oxygen states at metal surfaces, with XPS

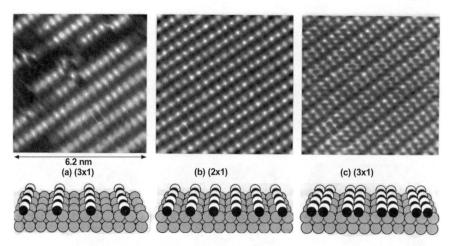

Figure 4.5 STM images (6.2 × 6.5 nm) observed in the chemisorption of oxygen at Ni(110) at room temperature (a) the (3 × 1)O state at θ = 0.33; (b) the (2 × 1)O state at θ = 0.5; (c) the (3 × 1)O state at θ = 0.66. Corresponding ball models of these are shown in (d), (e) and (f) and are typical of oxygen-induced reconstructions at metal surfaces. The small black balls represent the O adatoms. (Reproduced from Ref. 12).

establishing[2,15] that oxygen chemisorbed at magnesium, nickel and aluminium surfaces at 80 K were active for the oxidation of ammonia, water and carbon monoxide, respectively. These oxygen states were designated as $O^{\delta-}$ and precursors of, by comparison, the catalytically unreactive O^{2-} state.

Studies of *coadsorption* at Cu(110) and Zn(0001) where a coadsorbate, ammonia, acted as a probe of a reactive oxygen transient let to the development of the model where the kinetically "hot" $O^{\delta-}$ transient [in the case of Cu(110)] and the molecular $O_2^{\delta-}$ transient [in the case of Zn(0001)] participated in oxidation catalysis[16] (see Chapters 2 and 5). At Zn(0001) dissociation of oxygen is "slow" and the molecular precursor forms an ammonia–dioxygen complex, the concentration of which increases with decreasing temperature and at a reaction rate which is inversely dependent on temperature. Which transient, atomic or molecular, is significant in chemical reactivity is metal dependent.

By studying the reactivity of *preadsorbed oxygen* at various coverages at Cu(110) and examining a Monte Carlo simulation of the distribution of oxygen adatoms at a metal surface (Figure 4.6), it was concluded that only at very low coverages (θ = 0.01) was the oxygen present as isolated atoms and reactive.[17] For θ = 0.1, the majority of the oxygens were unreactive and present as clusters of 3–4 atoms. This conclusion, albeit based on some simple assumptions in the Monte Carlo simulations, was in agreement with the STM results of Brune *et al.*[7] for oxygen chemisorption at Al(111) (Figure 4.7) and provided the impetus to explore the chemical reactivity of oxygen states at metal surfaces by STM.

The classical picture, where following the transition state the two oxygen atoms are channelled downwards to the nearest available metal atoms, is

Dynamics of Surface Reactions and Oxygen Chemisorption

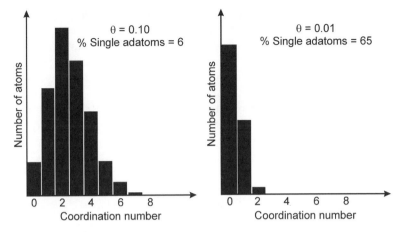

Figure 4.6 Monte Carlo simulation of the structure of oxygen chemisorbed at a metal surface; only at very low coverages is the oxygen present as isolated adatoms; at $\theta = 0.1$ the majority are present as clusters.

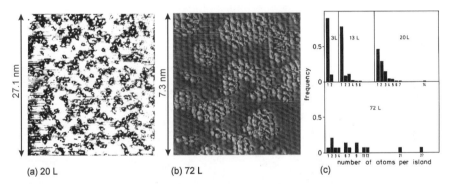

Figure 4.7 Images of oxygen chemisorption at Al(111) at room temperature: (a) 20 L exposure, (b) 72 L exposure and (c) frequency of island structures and number of atoms per island. (Reproduced from Ref. 7).

clearly not correct and that part of the energy is released as kinetic energy parallel to the surface and giving rise to translational motion of (at least) one of the oxygens. The lifetime of the oxygen surface transient at Al(111) was estimated to be of the order of 1 ps at 300 K, but this clearly depends on the coverage. Histograms of the relative proportions of various island sizes as a function of oxygen exposure and coverage indicate that at very low coverages single oxygen adatoms prevail, but at higher coverages collision with thermalised immobile oxygen adatoms leads to nucleation and growth of (1 × 1) islands. The conclusions of Brune et al.[7] were challenged by Schmid et al.[18] at Vienna in 2001, suggesting that the oxygen adatoms undergoing a transient motion of the order of 80 Å would be "rather astonishing" if correct. Schmid's group studied oxygen adsorption at Al(111) in the temperature range 130–195 K, then cooled to 80 K when the images were recorded. Under these conditions, pairs of oxygen adatoms were observed at distances from each other of between one and three aluminium substrate atom spacings. The surface was then annealed to "approximately 250 K" when there is evidence for greater pair separation – up to four aluminium spacings. These experiments are, of course, completely different to those reported by Brune *et al.* at 295 K and a strict comparison is not possible. What is obvious from the Fritz Haber Institute data is that following dissociation at 295 K, oxygen adatoms undergo surface hopping (diffusion), have a significant surface lifetime, are O^- like in character and will exhibit special chemical reactivity under catalytic oxidation conditions.

The oxidation of magnesium,[19] whether by the dissociative chemisorption of oxygen or nitrous oxide, exhibited some similarities to aluminium. At 295 K, the oxygen adatoms are kinetically hot and mobile, but nucleate to form hexagonal structures which are typically 0.3 nm in height. Although most of these structures are kinetically stable (immobile), there is evidence for some of them to be intrinsically unstable, with the upper layer undergoing translational motion relative to the lower layer across the Mg(0001) surface (Figure 4.8).

Line profiles of these structures indicate a step-height of between 0.14 and 0.15 nm for the overlapping Mg(0001)–O–Mg bilayer (Figure 4.9). Clearly, at

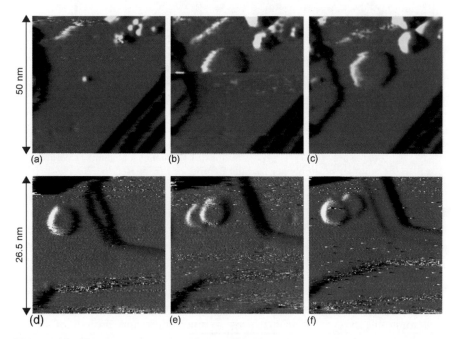

Figure 4.8 Oxygen adatom mobility resulting in the growth of oxide nuclei at Mg(0001) at 295 K (a–c); separation of "oxide bilayer" at Mg(0001) at 295 K (d–f). (Reproduced from Ref. 41).

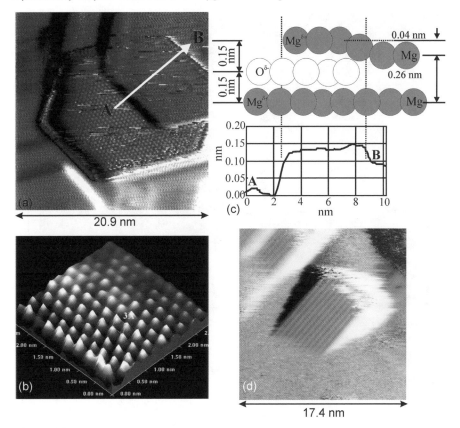

Figure 4.9 Oxygen chemisorption at Mg(0001) at 290 K: (a) 20.9 × 20.9 nm image of a (1 × 1)O adlayer partially overlayed by Mg atoms; (b) 3D image of the (1 × 1)O overlayer; (c) the relative height profile along the line A–B in (a) and a model of the step region; (d) a rectangular (square) oxygen state at the surface. (Reproduced from Ref. 19).

the interface the oxygen sites are "special" in that they are bonded at a step-edge involving the oxide–metal interface.

At high oxygen exposures at 295 K, the surface consists predominantly of hexagonal structures, but also present as a minor component are square lattice structures (Figure 4.10) reminiscent of the cubic structure associated with MgO "smoke" formed by the oxidation of magnesium at high temperature.[20] Therefore, two pseudomorphic "oxide" overlayers form at Mg(0001) at room temperature, but what factors control their separate growth are not known.

To probe the early stage of oxygen chemisorption, that is, prior to the onset of surface reconstruction and oxide formation and relevant to our coadsorption reactivity studies, there were obvious advantages for STM observations to be made at cryogenic temperatures.

In 1999, we reported[21] low-temperature studies of oxygen states at Cu(110). At 110 K the oxygen state present at high coverage is largely disordered but

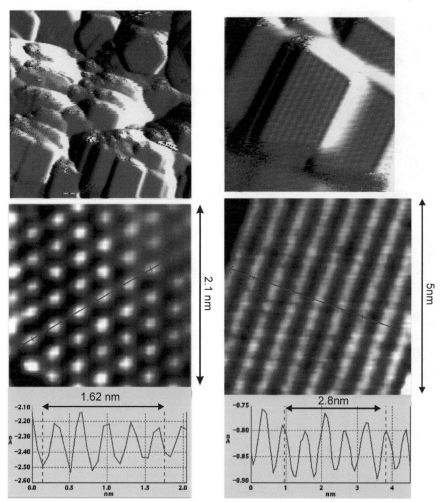

Figure 4.10 With increasing oxygen exposure at 295 K, the Mg(0001) surface consists of both hexagonal and square lattice structures; the line profiles indicate repeat distances of 0.321 and 0.56 nm in the atom resolved hexagonal and square structures, respectively, the former being the most prevalent structure present. (Reproduced from Ref. 41).

with some evidence for ordered structures running in the <100> direction (Figure 4.11). Each "site" imaged has an approximate dimension of 5 Å and is assigned to a molecularly adsorbed state. At 120 K at low coverage there is evidence for clustering at step-edges, the images are fuzzy but also with evidence for individual oxygen states present elsewhere on the surface. The oxygens within these clusters, which are separated from each other by up to 8 Å, are about 2 Å in diameter and assigned to oxygen adatoms $O^{\delta-}$. This is clear evidence for dissociation having taken place at 120 K.

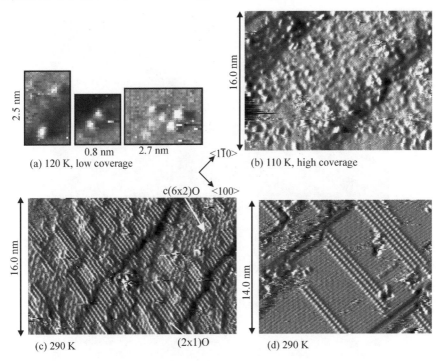

Figure 4.11 STM images of oxygen chemisorption at Cu(110): (a) at low coverage at 120 K; (b) at high coverage at 110 K; (c) (2 × 1) and c(6 × 2) oxygen states present after warming from 110 to 290 K; (d) (2 × 1)O strings present when oxygen is chemisorbed at 290 K. These distinct oxygen states would be expected to exhibit variations in chemical reactivity. (Reproduced from Ref. 21).

On warming to 300 K, the adlayer undergoes a disorder–order transition; the $O^{\delta-}$ states present at 120 K, together with the surface copper atoms, are highly mobile and can be considered to resemble a two-dimensional gas which at 300 K transforms into a structurally well-ordered immobile "oxide" adlayer.[22] This is very similar to the model proposed from spectroscopic (XPS) studies and based on chemical reactivity evidence (see Chapter 2).

At very low temperatures (4 K), Bradshaw and co-workers[23] observed with Cu(110) weakly bound, "trapped" oxygen molecules coexisting with pairs of atoms (Figure 4.12), both adsorbed in hollow sites. Oxygen molecules can therefore either be trapped in a local minimum of the potential energy surface or find a channel for dissociation. At such low temperatures, adsorption is immobile, no thermal diffusion being observed during the time-scale of the experiment.

In 2003, Salmeron and co-workers[24] reported STM images of oxygen states at Pd(111) at very low temperatures (25–210 K), providing a detailed picture of the surface site occupied by molecular oxygen (Figure 4.13). The dominant site is the fcc hollow site and the oxygen state is suggested to be the transient

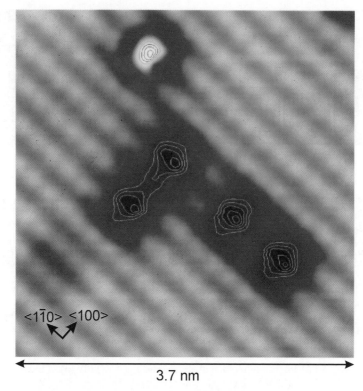

Figure 4.12 Oxygen chemisorption at Cu(110) at 4 K. The rows running in the [110] direction are atomically resolved copper atoms. (Reproduced from Ref. 23).

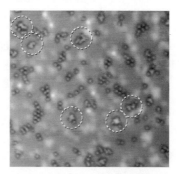

Figure 4.13 High-quality STM images of O_2 distributed at Pd(111) at 50 K. Small clusters are formed that exhibit (2×2) ordering, although more dense structures (indicated by circles) are also present. The inhomogeneous background is due to sub-surface impurities at a concentration of 0.03 monolayers. (Reproduced from Ref. 24).

peroxide-like O^-_2 state characterised by a vibrational frequency at $835\,cm^{-1}$. They appear as slightly elongated 5 Å protrusions. Above 100 K there is evidence for p(2 × 2) oxygen islands forming with dissociation occurring at 120 K from the periphery of island edges and also near sub-surface impurities.

Chemisorption of oxygen at Pt(111) has been studied in detail by Ertl's group[25] and the STM evidence is for complex structural features present in the temperature range 54–160 K (Figure 4.14). The limitations of the Langmuir model, frequently invoked for reactions at platinum surfaces, is obvious from

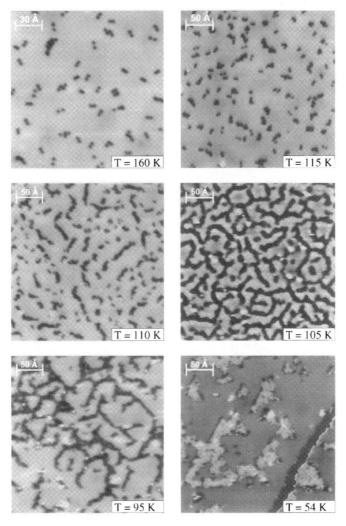

Figure 4.14 STM images of 1 L of oxygen exposed to Pt(111) at the temperatures indicated, emphasising the anisotropic growth of oxygen islands. The scale bar of the image at 160 K is 30 Å; that at all other temperatures is 50 Å. (Reproduced from Ref. 25).

the atom resolved images observed. At 160 K the adatoms which are randomly distributed across the surface always appear as pairs, the separation within the pairs being twice the lattice constant of the Pt(111) surface, 0.5–0.6 nm; there is no evidence for preferential adsorption at atomic steps.

4.4 Control of Oxygen States by Coadsorbates

The presence of coadsorbates can also control oxygen surface structures. Chemisorbed oxygen states at Cu(110) with short (2×1) strings can be "synthesised" by either chemisorption replacement reactions – exposing a presorbed oxygen adlayer to hydrogen sulfide – or by exposing a mobile, structurally disordered sulfur adlayer to oxygen.[26] The biphasic $c(2 \times 2)$S and (2×1)O states present reflect the lateral interactions involved, their surface distribution being kinetically controlled (Figure 4.15).

At a Pt(111) surface and a sulfur coverage of 0.25, the structure is a $p(2 \times 2)$ overlayer.[27] However, on coadsorbing carbon monoxide, structural reordering occurs, the surface structure being compressed into an ordered $(\sqrt{3} \times \sqrt{3})R30°$ state of higher local coverage, creating space on the surface for CO adsorption. New terraces form, containing exclusively carbon monoxide and separated

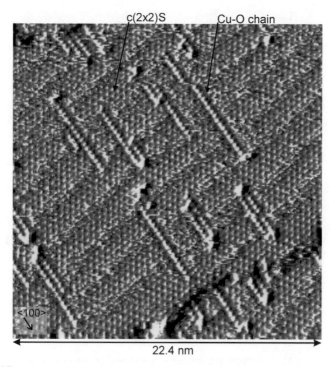

Figure 4.15 Chemisorptive replacement of oxygen at Cu(110) by sulfur resulting in a $c(2 \times 2)$S structure and isolated "unreactive" (2×1)O strings; Cu(110)–O + H$_2$S(g) → Cu(110)–S + H$_2$O(g). (Reproduced from Ref. 26).

from the terraces containing the compressed sulfur adlayer. Apparently the CO molecules are not observed by STM due to their high surface mobility. On heating, the CO desorbs and the original p(2 × 2)S adlayer re-forms.

4.5 Adsorbate Interactions, Mobility and Residence Times

We emphasised earlier in Chapter 2 how in the classical approach to surface dynamics the concept of site and surface residence times is essential to the discussion of adsorption phenomena, particularly when transient states participate in the reactions. Although LEED provided significant clues as to the role of adsorbate interactions in determining surface structures, with island sizes being estimated from the analysis of spot profiles, it was the development of fast STM that provided atom resolved information on the dynamics of oxygen chemisorption. The Ru(0001)–oxygen system investigated by Wintterlin[28] provides an outstanding contribution to our understanding of the dynamics of oxygen chemisorption at 300 K, while cryogenic studies of the Ag(110)– and Cu (110)–oxygen systems[22,29] revealed the role that dioxygen states play and also the facile nature of disorder–order transitions within the oxygen adlayer. Wintterlin et al.[30] drew attention in 1997 to the significant advantages of recording images in the "constant height" mode and at a fast imaging rate, 15 frames s^{-1}. At low oxygen coverages individual oxygen adatoms are imaged with occasional formation of dimers and trimers (Figure 4.16). By contrast, when imaged in the "constant current" mode, single atoms are not imaged but are seen as "streaks", the oxygen adatom having jumped during the time the tip takes to return to the atom and image it again. A jump rate of 14±3 s^{-1} is estimated for atoms which have no neighbours within distances of $\sqrt{7}$ lattice constants, jumps occurring with equal probability in all directions, suggesting that "tip effects" play no part in the dynamics observed.

When two or more oxygen adatoms are within two lattice constants apart, their residence times become longer than that characteristic of isolated atoms. This suggests that there are attractive interactions but are of the order of kT since the oxygen "clusters" are not stable and dissociate during scanning. There was no evidence for oxygen adatoms occupying nearest neighbour (1 × 1) sites; at higher coverage, islands with a (2 × 2) structure develop. By recording video sequences, "real time" movies of the surface dynamics are observed; the majority of atom jumps are between neighbouring sites although longer jumps are not ruled out (Figure 4.17). The observed jump rate at 300 K of $14 \pm 3 \, s^{-1}$, together with an assumed pre-exponential factor of 10^{13} s^{-1}, gives an estimated activation energy barrier to "hopping" of about 60 kJ mol^{-1}.

What became evident was that interactions between adsorbed particles can also exert an influence on their surface mobility and therefore the residence time at a particular site. The mean residence time of an isolated oxygen adatom at the Ru(0001) surface varies from 60 to 220 ms when a second oxygen adatom is located two lattice constants a_0 apart from the first but only 13 ms when the

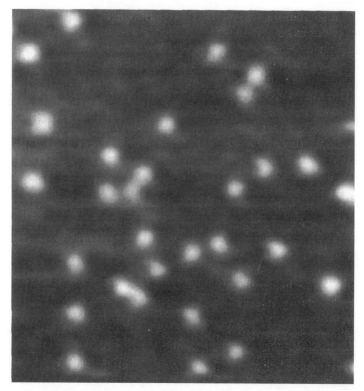

Figure 4.16 Constant-height STM image of Ru(0001), 15 frames s^{-1} with individual oxygen adatoms present. (Reproduced from Ref. 30).

separation distance is only $\sqrt{3}a_0$. Ertl[31] suggests that defining a single particle diffusion coefficient in the context of a surface reaction is not possible; the diffusion of adsorbates represents a highly complex phenomenon which makes meaningful kinetic studies very difficult. If, however, the adsorbed particles are sufficiently far apart for them not to be influenced by each other, then STM can provide useful diffusion data for the independent motion of single particles and Wintterlin summarised such information.

A series of more than 1000 successive images were taken of a pair of oxygen adatoms separated from all others; the two atoms underwent surface hopping between successive images and the distances were analysed statistically. The pathway followed by the hopping adatom at the Ru(0001) surface was shown to be influenced by the second oxygen atom, with the nearest site never occupied due to a strong repulsive interaction. Residence times differed from that of an isolated oxygen adatom, with each atom influencing 36 sites around it and times differing by more than an order of magnitude. By analysing such data, the interaction potential between two oxygen adatoms up to three lattice constant distances apart was computed. It has the expected shape, exhibiting an attractive and repulsive part to the curve with the minimum at $2a_0$ corresponding to the (2 × 2) structure observed at Ru(0001).

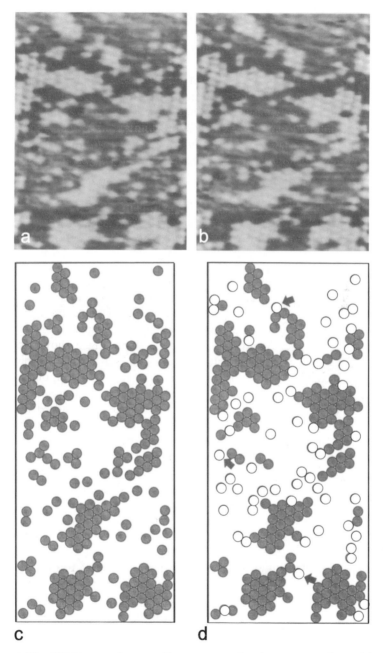

Figure 4.17 STM images from a video sequence showing oxygen adatom islands at Ru(0001) for a coverage of 0.09. (a) t = 0; (b) t = 0.17 s with the positions of the atoms illustrated in (c) and (d). Atoms marked brighter in (d) have moved with respect to those in (c); the arrows indicate examples of atom motions by which the size and shape of islands alter. (Reproduced from Ref. 30).

At Ag(110) oxygen molecules were observed by Barth et al.[32] undergoing surface hopping, coming to rest as ensembles or clusters, a consequence of undergoing collision with a second oxygen molecule. At a surface coverage of 0.02 and a temperature of 65 K, it was established by STM that the majority of the oxygen molecules were present as stable ensembles of two or four molecules (Figure 4.18). A statistical analysis showed that about 40% of the molecules at this coverage were ensembles whereas a random distribution would have predicted only 0.6%. This is direct experimental evidence for what were described as intrinsic precursor or trapping states central to the formulation of models for kinetic processes at metal surfaces and discussed in detail elsewhere (see Chapter 2).

With Ag(100), where oxygen is known to be dissociatively chemisorbed at 140 K, Schintke et al.[33] observed by STM that at very low oxygen coverages of 0.1–to 1%, two main interpair distances, 2.0 and 4.0 nm, were observed. These correspond to about seven and 14 lattice constants within a pair (Figure 4.19). The authors ruled out thermal motion as being responsible for these separations and concluded that they are the result of the dissociation event itself. Calculations of the migration distances of hot oxygen adatoms at Ag (100) by Zeiri[34] have shown that an energy release of 1.3 eV per atom would lead to an interpair distance of 2 nm. It is suggested that at Ag(100) the larger distance (4.0 nm) arises from oxygen molecules that have dissociated directly from the gas phase and the energy dissipated equally between the two oxygen atoms. The smaller interpair distance (2 nm) is the result of dissociation occurring from a molecular precursor.

It was real-time XPS studies of oxygen chemisorption at magnesium, aluminium and copper that drew attention to the possible role of oxygen transients

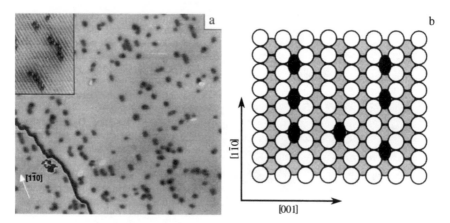

Figure 4.18 (a) STM image (39 × 23 nm) O_2 molecules at Ag(110) at 65 K, illustrating the hot precursor mechanism at a coverage of 0.02. The inset shows an atomic resolution image of the silver surface and the O_2 molecules as dark holes. Also shown (b) is a ball model with oxygen molecules (black) and surface silver atoms (white) and second layer silver atoms (grey). (Reproduced from Ref. 32).

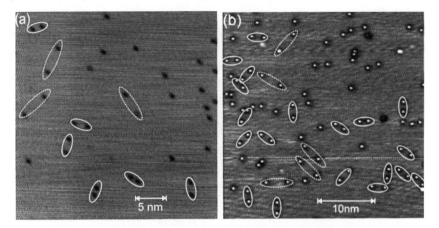

Figure 4.19 Pairing of oxygen adatoms on Ag(100) at 140 K: (a) oxygen coverage 0.13% ML; (b) oxygen coverage 0.5% ML. (Reproduced from Ref. 33).

undergoing surface diffusion and exhibiting chemistry closely associated with the O^- state, with Ertl providing conclusive STM evidence[7] for oxygen mobility at Al(111) at 300 K. Subsequently, and with the availability of low-temperature STM, the events accompanying oxygen chemisorption were shown to be complex and temperature dependent. At Pt(111), the oxygen adatoms at 160 K were paired and randomly distributed[25] over the surface (Figure 4.14). However, at 105 K, the atom pairs were arranged in chains, with Zambelli *et al.*[25] concluding that the dissociation probability of the molecular precursor is enhanced when adjacent to oxygen adatoms at the ends of chains, leading to chain growth. At 105 K, the surface lifetime prior to desorption is sufficiently long for the molecular precursor to locate an oxygen adatom before it is desorbed and therefore result in chain growth (Figure 4.20).

4.6 Atom-tracking STM

Although there is the view that the conventional STM image acquisition is limited for studying the diffusion of adatoms by the rate at which dynamic events can be resolved, cryogenic studies, by slowing the process, have provided a way forward. This was the approach adopted by the Fritz Haber Institute group. Data did exist, however, from field emission and field ion microscopy, the latter being the first experimental method able of resolving individual atoms on solid surfaces. This was due to Muller's group in the mid-1950s, with quantitative data becoming available somewhat later, mainly for metal atom diffusion on metals from the groups of Ehrlich[35] at General Electric at Schenectady, USA, and Bassett[36] at Imperial College, London. Two mechanisms for surface diffusion have now emerged, the more conventional "hopping process" and an "exchange mechanism" where the diffusing atom buries itself in the surface and pushes an atom out to take its place.

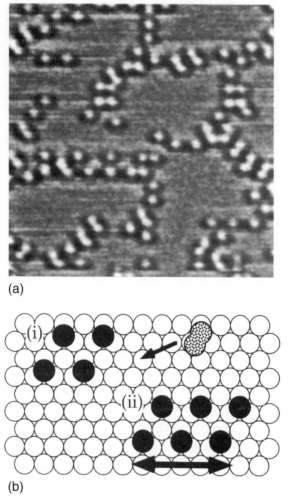

Figure 4.20 (a) Chains of oxygen adatoms formed by oxygen chemisorption at Pt(111) at 105 K (70 × 70 Å). (b) Model illustrating the growth of O chains by the collision of precursor molecular oxygen with the ends of O chains where they dissociate. (Reproduced from Ref. 28).

Much effort is being put into developing general rules that can be applied to describe and predict how atoms and molecules diffuse at solid surfaces. The direct measurement of diffusion using atom-tracking STM is an approach that has considerable potential. This novel development, due to Swartzentruber[37] at Sandia Laboratories in 1996, can resolve every diffusion event with data being reported for silicon dimers at Si(001) at temperatures between 295 and 400 K. The tip is locked on to the dimer with sub-ångstrom precision and tracks the dimer as it diffuses across the surface. The sensitivity of the method, according to Swarzentruber, to follow dynamic events is increased by a factor of nearly 1000 over the conventional STM imaging technique. The mean residence time τ (time between "hops") of the dimers is 0.11 s at 400 K, with evidence that over the temperature range studied surface diffusion is an activated process obeying the relationship

$$\frac{1}{\tau} = \nu_0 \exp\left(-\frac{E_a}{RT}\right) \qquad (3)$$

where the activation energy E_a is 0.94 eV and $\nu_0 = 10^{12.8}$ s^{-1}, the latter being close to the value of 10^{13} s^{-1} often assumed for surface diffusion. As far as we are aware, atom-tracking STM has not been used to study molecular events associated with surface chemistry or catalysis. Its advantages seem obvious.

4.7 Hot Oxygen Adatoms: How are they Formed?

Although the STM evidence is that abstractive chemisorption of oxygen leads to single oxygen adatoms, with the second oxygen atom being well separated from the first, some of the adsorption energy could also result in oxygen atoms (ions) being desorbed. Gas-phase oxygen ions have been observed with alkali metals.[38]

In view of the doubts expressed by Schmid et al.[39] regarding the existence of well-separated oxygen adatoms (up to 8 nm) resulting from the dissociation of a single molecule, Binetti and Hasselbrink[40] have more recently established that at Al(111), using molecular beams and laser spectroscopy, at low coverages one oxygen adatom forms and the second appears in the gas phase (Figure 4.21). Abstractive chemisorption operates at all translational energies but increases

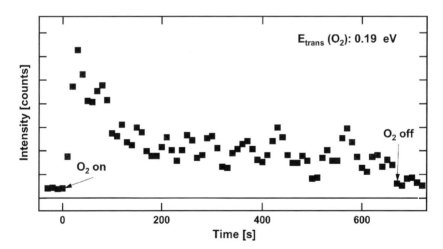

Figure 4.21 Evidence for the hot atom concept from ionisation laser spectroscopy. Time-dependent evolution of the gas-phase O atom signal at $E_{trans} = 0.19$ eV. At $t=0$ the monitoring starts and is completed at $t=600$ s when the oxygen surface coverage at the Al(111) surface is 0.2. (Reproduced from Ref. 40).

with energy while rotational excitation of the molecules suppresses the abstraction process. The gas-phase oxygen atoms are shown to possess translational energy equivalent to about one-seventh of the adsorption energy. For a translational energy of 0.19 eV and up to a coverage of 0.2, the concentration of oxygen atoms (in the gas phase) is at a maximum within the first 50 s of the exposure time and then decreases rapidly over the next 150 s. The authors suggest that the low translational energy might be a consequence of the concerted motion of the incoming oxygen molecule and the substrate metal atoms on the formation of the Al–O bond, while the O–O bond length increases, resulting in energy being released to many degrees of freedom. Furthermore, the recoiling oxygen atom, which initially has the character of an O^- ion, is suggested to lose the electron on the outgoing trajectory. This means that the charge-transfer process opens up channels for the dissipation of energy into electronic degrees of freedom of the metal substrate.

What is also interesting is that the O^- ion is the state which is compatible with the mechanistic deductions regarding the active oxygen in coadsorption oxidation studies.[1,2,3,13] What also follows is that in modelling catalytic oxidation reactions – as for example the oxidation of propene at Mg(0001) – account should be taken as to whether the oxygen atoms (ions) that appear in the gas phase have a finite surface lifetime before desorbing. This would then conform to the essential features of what has been described as a "two-dimensional" gas reaction[41] (see also Chapter 2). What also stands out is that the catalytic oxidation (coadsorption) studies at copper, magnesium and aluminium were of reactions in which oxygen chemisorption is a highly exothermic process favouring energy being partitioned to provide surface translation or/and desorption of oxygen "atoms".

Scheffler and colleagues at the Fritz Haber Institute[42] have recently drawn attention to the possible significance of spin selection rules for explaining the low sticking probability of dioxygen dissociative chemisorption at Al(111). These studies show that the adsorption energy is "efficiently transferred to strong surface vibrations and that the oxygen adatoms do not move far". They approach the problem from recognising that chemical interactions are ruled by various selection rules and that in this case spin conservation is expected to be applicable. They conclude that when O_2 approaches the Al(111) surface oriented perpendicular to the surface the spin is shifted to the atom that is further away from the surface. This is suggested to give rise to the abstractive chemisorption process, one oxygen atom adsorbing in the singlet state with the spin being efficiently carried away with the other oxygen atom, which is either ejected into the gas phase or trajected along the surface "to some distant place". The authors reject the hot atom concept for explaining the experimental observations.

4.8 Summary

Spectroscopic studies during the period 1986–1990 drew attention through coadsorption studies to transient oxygen states existing when a dioxygen molecule adsorbed at a metal surface underwent dissociative chemisorption (see Chapter 2). During the decade 1992–2002, the understanding of the dynamics of oxygen chemisorption at metal surfaces took a significant step forward through the availability of STM. Although much of the early data were confined to room temperature, due to instrumental limitations, the advantages of studies at cryogenic temperatures for following details of the atomic events prior and subsequent to bond cleavage became clear.

In 1992, the Fritz Haber Institute group first reported that for oxygen dissociative chemisorption at Al(111) at 300 K the two oxygen adatoms were separated by at least 80 Å before being accommodated at the surface.[7] The classical view that the two oxygen adatoms were bonded at adjacent sites was clearly untenable. The authors estimated that the "translational lifetime" of the oxygen transient was 1 ps, which is of the same order of magnitude of that estimated[2] for the O^- state at Mg(0001) surface before it became deactivated in the "oxide-like" 2e state O^{2-}(a). Lifetimes of transients are, however, dependent on surface coverage and temperature. Support for the hot atom concept also came from the detection of O^- species in the *gas phase* during chemisorption at caesium surfaces and which Greber *et al.*[38] attributed to electronic excitation originating from the exothermicity of the reaction (the probability of this event occurring was, however, very low). Scheffler *et al.* took a different view. However, in 2004, Binetti and Hasselbrink provided experimental evidence[40] for abstractive chemisorption of oxygen at Al(111), with the recoiling gas-phase oxygen atom having initially the character of an O^- ion.

It was STM studies at cryogenic temperatures, however, that were the key experiments in drawing attention to the unusual structural and kinetic behaviour accompanying oxygen adsorption and dissociation at metal surfaces. The "skating" of oxygen molecules following adsorption at Ag(110) at a very low temperature, 60 K, indicated that thermally induced diffusion was not involved with the molecules coming to rest due to collision with other oxygen molecules. Oxygen clusters were observed at Cu(110) even at 4 K with a variety of states in the temperature range 100–295 K with the stable reconstructed (2 × 1) state present at 295 K. The separation of oxygen adatoms following dissociative chemisorption were usually much smaller than that observed with Al(111). There was also evidence that on Pt(111) dissociation could also be enhanced when a molecule was in close proximity to an oxygen adatom, resulting in chain-like structures being formed at low temperatures (105 K). Real-time evidence emphasised the chaotic behaviour of oxygen chemisorption, the difficulty of developing meaningful kinetic expressions and the significance of attractive interactions which favoured island growth. In the next chapter, we shall consider how these structures and the dynamics of the surface processes can influence the chemistry of oxidation catalysis, the design of new catalysts and also relate to the models developed from surface spectroscopies discussed in Chapter 2. What is also clear is that the dynamics of a catalytic reaction cannot be deduced from studies of the reacting molecules *separately*.

References

1. M. W. Roberts, *J. Chem. Soc., Faraday Trans. 1*, 1987, **87**, General Discussion, 2085 (Faraday Symposium No. 21).
2. C. T. Au and M. W. Roberts, *J. Chem. Soc., Faraday Trans. 1*, 1987, **83**, 2047; A. F. Carley and M. W. Roberts, *J. Chem. Soc., Chem. Commun.*, 1987, 355.
3. C. T. Au, M. W. Roberts and A. R. Zhu, *J. Chem. Soc., Chem. Commun.*, 1984, 737.
4. T. Matsushima, *Surf. Sci.*, 1983, **127**, 403.
5. C. B. Mullins, C. T. Rettner and D. J. Auerbach, *J. Chem. Phys.*, 1991, **95**, 8649.
6. M.E. Davis, R.A. Van Santen and H. Niemansverdriet, Cattech Kluwer/Academic, New York, 1997, 63.
7. G. Ertl, *Top. Catal.*, 1994, **1**, 305; H. Brune, J. Wintterlin, J. Trost, G. Ertl, J. Wiechers and R. J. Behm, *J. Chem. Phys.*, 1993, **99**, 1993; H. Brune, J. Wintterlin, R. J. Behm and G. Ertl, *Phys. Rev. Lett.*, 1992, **68**, 624.
8. See, for example, M. W. Roberts and C. S. McKee, *Chemistry of the Metal–Gas Interface*, Clarendon Press, Oxford, 1978, p. 439.
9. M. W. Roberts and B. R. Wells, *Discuss. Faraday Soc.*, 1966, **41**, 162.
10. D. J. Coulman, J. Wintterlin, R. J. Behm and G. Ertl, *Phys. Rev. Lett.*, 1990, **64**, 1761.
11. A. F. Carley, P. R. Davies and M. W. Roberts, unpublished work.
12. L. Eirdal, F. Besenbacher, E. Laesgsgaard and I. Stensgaard, *Surf. Sci.*, 1994, **312**, 31.
13. M. W. Roberts, *Chem. Soc. Rev.*, 1989, **1**, 451 *Surf. Sci.*, 1994, **299/300**, 769.
14. F. Besenbacher and I. Stensgaard, in *The Chemical Physics of Solid Surfaces*, Vol. 7, ed. D. A. King and D. P. Woodruff, Elsevier, Amsterdam, 1994, 573.
15. G. U. Kulkarni, C. N. R. Rao and M. W. Roberts, *Langmuir*, 1995, **11**, 2572.
16. Reviewed in ref. 13; see also Chapter 2.
17. A. F. Carley, P. R. Davies, M. W. Roberts and D. Vincent, *Top. Catal.*, 1994, **1**, 35.
18. M. Schmid, G. Leonardelli, R. Tscheliessnig, A. Bierdermann and P. Varga, *Surf. Sci.*, 2001, **478**, L355.
19. A. F. Carley, P. R. Davies, K. R. Harikumar, R. V. Jones and M. W. Roberts, *Top. Catal.*, 2003, **24**, 51 *Chem. Commun.*, 2002, 2021.
20. C. F. Jones, R. A. Reeve, R. Rigg, R. L. Segall, R. St. C. Smart and P. S. Turner, *J. Chem. Soc., Faraday Trans. 1*, 1984, **80**, 2609.
21. A. F. Carley, P. R. Davies, G. U. Kulkarni and M. W. Roberts, *Catal. Lett.*, 1999, **58**, 33.
22. A. F. Carley, P. R. Davies, R. V. Jones, K. R. Harikumer, G. U. Kulkarni and M. W. Roberts, *Top. Catal.*, 2000, **11/12**, 299; A. F. Carley, P. R. Davies and M. W. Roberts, *Catal. Lett.*, 2002, **80**, 25.

23. B. G. Briner, M. Doering, H.-P. Rust and A. M. Bradshaw, *Phys. Rev. Lett.*, 1997, **78**, 1516.
24. M. K. Rose, A. Borg, J. C. Dunphy, T. Mitsui, D. F. Ogletree and M. Salmeron, *Surf. Sci.*, 2003, **547**, 162.
25. T. Zambelli, J. V. Barth, J. Wintterlin and G. Ertl, *Nature*, 1997, **390**, 495; J. Wintterlin, R. Schuster and G. Ertl, *Phys. Rev. Lett.*, 1996, **77**, 123.
26. A. F. Carley, P. R. Davies, R. V. Jones, K. R. Harikumar, G. U. Kulkarni and M. W. Roberts, *J. Chem. Soc., Soc., Chem. Commun.*, 2000, 185.
27. M. Salmeron and J. Dunphy, *Faraday Discuss.*, 1996, **105**, 151.
28. J. Wintterlin, *Adv. Catal.*, 2000, 131.
29. J. V. Barth, T. Zambelli, J. Wintterlin and G. Ertl, *Chem. Phys. Lett.*, 1997, **270**, 152.
30. J. Wintterlin, J. Trost, S. Renisch, R. Schuster, T. Zambelli and G. Ertl, *Surf. Sci.*, 1997, **394**, 159.
31. G. Ertl, *Adv. Catal.*, 2000, **45**, 1.
32. J. Barth, T. Zambelli, J. Wintterlin and G. Ertl, *Chem. Phys. Lett.*, 1997, **270**, 152.
33. S. Schintke, S. Messerli, K. Morgenstern, J. Nieminem and W.-D. Schneider, *J. Chem. Phys.*, 2001, **114**, 4206.
34. Y. Zeiri, *J. Chem. Phys.*, 2000, **112**, 3408.
35. G. Ayrault and G. Ehrlich, *J. Chem. Phys.*, 1972, **57**, 1788.
36. D. W. Bassett, *Surface and Defect Properties of Solids*, Vol. 2, Chemical Society, London, 1973, p. 34.
37. B. S. Swartzentruber, *Phys. Rev. Lett.*, 1996, **76**, 459.
38. T. Greber, R. Grobecker, A. Morgante, A. Böttcher and G. Ertl, *Phys. Rev. Lett.*, 1993, **70**, 1331.
39. M. Schmid, G. Leonardelli, A. Tscheleissing, A. Biederman and P. Varga, *Surf. Sci.*, 2001, **478**, L355.
40. M. Binetti and E. Hasselbrink, *J. Phys. Chem.*, 2004, **108**, 14677.
41. A. F. Carley, P. R. Davies and M. W. Roberts, *Philos. Trans. R. Soc. London, Ser. A*, 2005, **363**, 829.
42. J. Behler, B. Delley, S. Lorenz, K. Reuter and M. Scheffler, *Phys. Rev. Lett.*, 2005, **94**, 036104–1.

Further Reading

G. Ertl, Elementary steps in heterogeneous catalysis, *Angew. Chem. Int. Ed.*, 1990, **29**, 1219.

E. K. Rideal, *Concepts in Catalysis*, Academic Press, New York, 1968.

G. A. Somorjai, *Principles of Surface Chemistry*, Prentice Hall, Englewood Cliffs, NJ, 1972.

R. J. Madix, Selected principles in surface reactivity: reaction kinetics on extended surfaces and the effects of reaction modifiers on surface reactivity, in *The Chemical Physics of Solid Surfaces and Heterogeneous Catalysis*, Vol. 4, ed. D. A. King and D. P. Woodruff, Elsevier, Amsterdam, 1982, 1.

R. A. van Santen and J. W. Niemantsverdriet, *Chemical Kinetics and Catalysis*, Plenum Press, New York, 1995.

D. J. Dwyer and F. M. Hoffmann (eds), *Surface Science of Catalysis: In situ Probes and Reaction Kinetics*, ACS Symposium Series, Vol. 482, American Chemical Society, Washington, DC, 1992.

K. Tanaka and M. Ikai, Adsorbed atoms and molecules destined for a reaction, *Top. Catal.*, 2002, **20**, 25.

G. A. Somorjai and Y. Borodko, Adsorbate (substrate)-induced restructuring of active transition metal sites of heterogeneous and enzyme catalysts, *Catal. Lett.*, 1999, **59**, 89.

T. Schalow, B. Brandt, D. E. Starr, M. Laurin, S. Schauermann, S. K. Shaikhutdinov, J. Libuda and H.-J. Freund, *Catal. Lett.*, 2006, **107**, 189.

CHAPTER 5
Catalytic Oxidation at Metal Surfaces: Atom Resolved Evidence

"We seek it here, we seek it there
In defects, steps most everywhere
Is it fact? Or merely a sleight
That deemed elusive active site"

Baroness Orczy

5.1 Introduction

Catalytic oxidation has been one of the most extensively studied areas in heterogeneous catalysis, with the activation of C–H, N–H, S–H and O–H bonds receiving particular attention.[1] Although single-crystal metal substrates have played a significant role in providing relationships between catalytic activity and structural features of the metal surface (*e.g.* the role of step sites, kinks, *etc.*), Sachtler drew attention[2] to the paucity of studies at single crystals which involved dynamic studies of gas mixtures – conditions close to those in "real catalysis". It was an aspect that we attempted to correct, first using surface spectroscopies[3] as discussed in Chapter 2, but followed up later by STM,[4] with the distinction being made between preadsorbed oxygen states and those present during dynamic (mixture) studies. We consider experimental data obtained using both approaches so as to illustrate the very different reactivity of oxygen states present in both cases and for which the quantum mechanical calculations of Neurock *et al.*[5] provided theoretical support. As a corollary, the coadsorption studies also provided the first evidence for transient oxygen states present during the dynamics of oxygen chemisorption at metal surfaces discussed in Chapter 4.

Spectroscopic studies (XPS and HREELS) established first in 1980 that the activity of oxygen states in the oxidation of ammonia at copper–O surfaces was

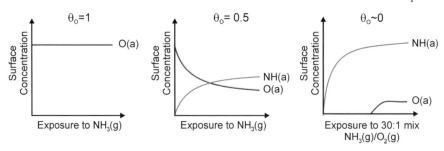

Figure 5.1 XPS evidence for oxygen states active in the oxidation of ammonia at Cu(110) at 290 K, for oxygen coverages of $\theta = 1.0$ and 0.5 and for an ammonia-rich NH_3–O_2 mixture. Note the high activity for NH formation with the 30:1 mixture.

dependent on the oxygen coverage at 290 K. The surface was inactive for $\theta = 1.0$ and showed limited activity for $\theta = 0.5$, but when the clean Cu(110) surface was exposed to an ammonia-rich NH_3–O_2 mixture a monolayer of NH(a) species was formed[6] at a "fast rate" at 290 K (Figure 5.1).

The fraction of oxygen adatoms reactive resulting in their chemisorptive replacement by NH_x species was determined from the quantification of the O(1s) and N(1s) spectra with HREELS providing further structural information on the nitrogen species. An STM study by Bradshaw's group provided good STM images of the structure of oxygen adatoms at Cu(110), and this was modelled successfully[7] with a Monte Carlo simulation using interaction energies of 2 and 7 kJ mol^{-1} in the [110] and [100] directions, respectively. The proportion of oxygen reactive to ammonia at different oxygen coverages, determined from the O(1s) and N(1s) spectra, was compared with the oxygen states that could be recognised in the Monte Carlo simulation of the images. Four different sites were considered – the centre of the oxygen islands, the [110] and [100] edges of the islands (the chain end sites) and the isolated oxygen adatoms. A strong correlation was found between the experimentally observed reactive oxygen adatoms and the total concentration of the oxygen adatoms at chain ends and the relatively small number that were isolated (Figure 5.2). What, then, have we learnt directly from STM studies of catalytic oxidation reactions: how do these modelling studies – largely based on spectroscopic information – provide a more general framework of oxidation reactivity at metal surfaces and what is the nature of the highly reactive oxygen state present in the NH_3–O_2 mixture experiment?

5.2 Ammonia Oxidation

At Cu(110) surfaces, a number of different oxygen states have been investigated by STM: (a) Cu(110)–O where the oxygen coverage is close to unity; (b) Cu(110)–O where the oxygen coverage is <1.0; and (c) Cu(110) exposed to an oxygen–ammonia mixture.

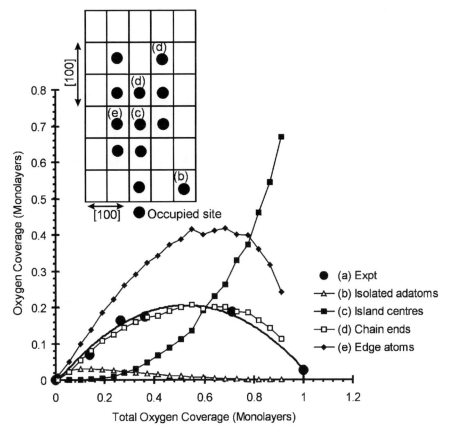

Figure 5.2 Oxygen states present at the ends of –Cu–O–Cu–O– chains are established as the "active sites" in ammonia oxidation at Cu(110) from a Monte Carlo simulation of the growth of the oxygen adlayer. The reactivity (the experimental curve) is best fitted to the atoms present at chain ends. (Reproduced from Ref. 7).

5.2.1 Cu(110) Pre-exposed to Oxygen

For an oxygen surface atom coverage of $5.1 \times 10^{14}\,\text{cm}^{-2}$ ($\theta \approx 1.0$), the surface is unreactive to ammonia at 295 K. However, on heating to 375 K reactivity is observed but limited to those oxygen states present at the ends of the (2×1) oxygen chains terminating at step edges.[4,8] NH species replace these oxygen atoms (water is desorbed) characterised by an N(1s) binding energy of 399 eV; the oxygen atoms within the close-packed (2×1) layer are unreactive (Figure 5.3a). The Monte Carlo simulation[7] of the reactivity observed supported the model that it was oxygens at the ends of chains and isolated oxygen adatoms that were the reactive sites.

At higher temperature (550 K), all the oxygen atoms are reactive and are replaced by nitrogen adatoms with characteristic structural features arranged in

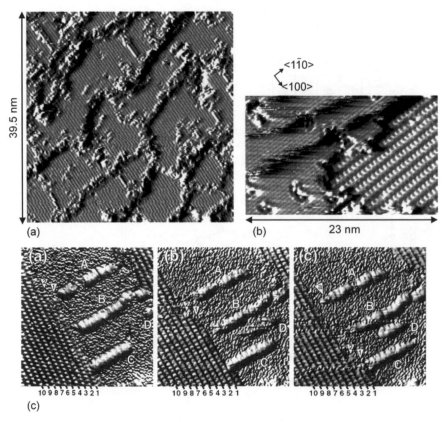

Figure 5.3 (a) STM established that at a Cu(110)–O overlayer ($\theta = 1.0$) at 375 K, the reactivity to ammonia to form NH(a) species (with water desorption) is confined to oxygens at the ends of chains terminating at step edges (Reproduced from Ref. 8). (b) At 550 K all oxygen adatoms are reactive and undergo a chemisorptive replacement reaction to give nitrogen adatoms arranged in a (2×3) configuration, running in the $\langle 1\bar{1}0 \rangle$ direction. (c) A sequence of STM images observed by Guo and Madix during oxydehydrogenation of ammonia at 300 K at a Cu(110)–O surface. The –Cu–O– rows are marked 1–10; perpendicular to these are imide species labelled A–D. Rows 1 and 2 decrease in length during exposure to NH_3, the oxygen desorbs as H_2O and imide species are chemisorbed. (Reproduced from Ref. 9).

a (2×3) configuration running in the <110> direction (Figure 5.3b). The N(1s) binding energy is at 397 eV, well established as indicative of chemisorbed nitrogen adatom N(a) and formed by the complete dehydrogenation of ammonia accompanied by water desorption.[4,8]

In 1996, Guo and Madix[9] at Stanford studied ammonia oxidation at Cu(110) pre-exposed to oxygen, observing the specific reactivity of oxygen states at the ends of (2×1) chains (Figure 5.3c) but with occasional evidence for activity within and at the edges of the (2×1)O islands. They suggested that the formation of NH species at the ends of chains inhibit reaction to be generated along the chain lengths. Although the conclusions from the Stanford and Cardiff groups were in general agreement, detailed comparisons are not possible. In the Stanford studies, Cu(110) was exposed to oxygen at 450 K, then to ammonia at 300 K followed by annealing at 410 K before taking STM images at 300 K; this is an inherently complex procedure which can influence the reactivity and the structural state of the final surface observed by STM at 300 K. The Cardiff group maintained the surface at a fixed temperature, 295 K, and also had the advantage of *in situ* XPS for the providing chemical information, essential for developing mechanisms. That step sites were special in surface reactivity was generally accepted, but Guo and Madix[10] established in some unique STM studies that the oxygen (2×1) state present at step sites on Cu(110) exhibited site-specific reactivity in the oxidation of ammonia. Reactivity was high at both the bottom and top of a $[1\bar{1}0]$ step and the bottom of an $[\bar{0}01]$ step whereas an oxygen site bonding to the top of an $[001]$ step was virtually inactive.

5.2.2 Coadsorption of Ammonia–Oxygen Mixtures at Cu(110)

Follow-up investigations of the earlier spectroscopic studies[6] were designed to simulate a catalytic reaction, albeit at low pressures, with both chemical and structural information available from XPS and STM, respectively. With a 30:1 ammonia-to-oxygen ratio imide strings were formed[11] at 290 K running in the <110> direction, *i.e.* at right-angles to the oxygen (when present) and along the copper rows (Figure 5.4). The separation between the "imide" rows is 0.72 nm, which is close to twice the copper lattice spacing (0.36 nm) in the <100> direction. Although the NH species are not resolved along the strings, quantification of the N(1s) intensity at 398 eV – well established as

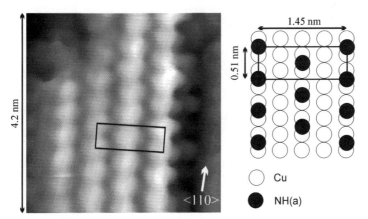

Figure 5.4 Coadsorption of a 30:1 ammonia–oxygen mixture at a Cu(110) surface at 290 K with the formation of well ordered $c(2 \times 4)$ imide chains running in the <110> direction. The separation between the rows is 7.2 Å and within the rows 5.1 Å, the NH species occupying the bridge sites. (Reproduced from Ref. 11,39).

characteristic of imide species – gave an NH concentration of $2.6 \times 10^{14}\,\text{cm}^{-2}$, which suggests that they are chemisorbed in alternate short bridged sites with an NH–NH spacing of ca 0.5 nm. Whether the imide species are present as a p(2 × 2) or a c(2 × 4) structure could not be established.

At higher temperature (475 K) and a similar ammonia-to-oxygen ratio, dehydrogenation of ammonia is complete, resulting in the chemisorption of nitrogen adatoms, characterised by N(1s) intensity at 397 eV and a (2 × 3) string structure running in the <110> direction.[8] There was no evidence for surface oxygen states being present in coadsorption studies at either 290 or 475 K (Figures 5.4a and 5.5b).

When a Cu(110) surface was exposed to a 1:1 ammonia–oxygen mixture at 60 K and the surface warmed to 290 K, the structural features observed indicated both nitrogen adatoms present as (2 × 3) strings running in the <110> direction and oxygen adatoms present in the characteristic (2 × 1)O strings running in the <100> direction[4,12] (Figure 5.5a). It is clear that the metastable oxygen states present at low temperatures (<120 K) and known to be disordered, are active in the complete dehydrogenation of ammonia. This high activity is both novel and unexpected. However, on warming the oxygen adlayer to 290 K there is competition between ordering of the disordered oxygen state, to give the unreactive (2 × 1) O strings and hydrogen abstraction by the disordered oxygen to generate chemisorbed nitrogen adatoms. At 290 K, there is therefore present a biphasic surface structure composed of (2 × 3)N and (2 × 1)O domains. An essential prerequisite for these structures to develop is a high mobility of both oxygen and copper atoms, features characteristic of the Cu(110)–O surface at low temperatures.

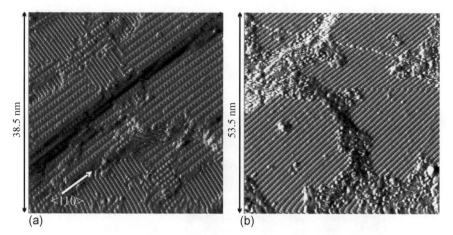

Figure 5.5 (a) Coadsorption of a 1:1 NH_3–O_2 mixture at 60 K followed by warming to 290 K; oxidation of ammonia is complete to give (2 × 3)N strings running in the <110> direction and (2 × 1)O rows running in the <100> direction (Reproduced from refs 8, 39). (b) At 475 K a 30:1 NH_3–O_2 mixture exposed to Cu(110) generated a complete monolayer of nitrogen adatoms in a (2 × 3) structure.

5.2.3 Coadsorption of Ammonia–Oxygen Mixtures at Mg(0001)

It was coadsorption studies at Mg(0001) by surface spectroscopies that first established in 1986 the role that oxygen transients $O^{\delta-}(s)$ had in the oxydehydrogenation of ammonia. Structural studies by STM[12] showed that under similar experimental conditions the Mg(0001) surface was transformed at 295 K to reveal a predominately hexagonal structure identical with that observed in oxygen chemisorption, with XPS indicating strong intensities in the O(1s) region at 530.5 eV and the N(1s) region at 399 eV. These are binding energies characteristic of chemisorbed oxygen and amide species and analysis of the spectra indicates atom concentrations of 5.3×10^{14} and $2.6 \times 10^{14}\,cm^{-2}$, respectively. The atom resolved hexagonal structure is epitaxial with the oxygen (1×1) adlayer, the line profile indicating a spacing of 0.321 nm matching closely the Mg–Mg distance in an Mg(0001) surface with a height difference of generally 0.125 nm between the hexagonal islands, but occasionally as much as 0.6 nm. No distinct structural features that can be attributed to nitrogen species were observed, but there were disordered areas of the surface. Since nitrogen states present at an Mg(0001) surface appear in general to be disordered, as for example in NO dissociation,[12] we associate the chemisorbed NH_2 species with this disorder.

5.2.4 Ni(110) Pre-exposed to Oxygen

Structural (STM) studies of oxygen chemisorption at Ni(110) indicate surface restructuring[13] with added –Ni–O– rows formed along the <100> direction at 295 K. The spacing between the rows decreases with increasing oxygen coverage and well-defined (3×1) and (2×1) phases are present (Figure 5.6). At high

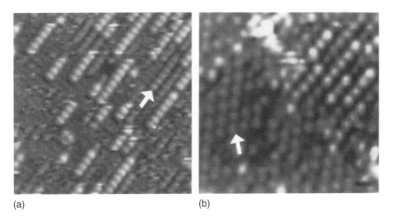

(a)   (b)

Figure 5.6 (a) STM image showing the appearance of OH molecules running in the <001> direction at an Ni(110)–O surface ($\theta_0 = 0.17$) after exposure to 0.2 L NH_3 at room temperature. With further exposure to ammonia (b), a $c(2 \times 2)$ structure develops (marked with an arrow) that is attributed to $NH_2(a)$. (Reproduced from Ref. 13).

oxygen coverage the surface is, as with Cu(110), unreactive to ammonia, but at low oxygen coverage dehydrogenation of ammonia is observed. This Ruan et al.[13] at Aarhus University ascribed to the activity of oxygen atoms which terminate the short, mobile –Ni–O– added rows, present at low oxygen coverage. The new structures observed in 1994 run in the same direction as the Cu(110)–O rows and were attributed to OH groups. The NH_x species were mobile and not structurally resolved at this stage, but with further exposure to ammonia the OH species were removed (as water) and the NH_x species observed with a c(2 × 2) structure. The added row Ni(110)–O structure present at this stage remained unreactive and immobile at 300 K; their inactivity was attributed to the rows not breaking up into smaller segments and exposing terminal oxygen atoms. There are obvious similarities with the model suggested for ammonia oxidation at Cu(110) and the authors attributed the abstraction of hydrogen to oxygen atoms that terminate the –Ni–O– rows leaving them terminated by nickel atoms. The hydroxyl species appear in the image as a different type of <100> directed row with small protrusions. These are very close to the short bridge sites and the rows form preferentially near the ends of –Ni–O– rows which simultaneously are reduced in number. A c(2 × 2) structure observed at high exposure to ammonia results in the formation of small domains of c(2 × 2) symmetry which the authors attributed to NH_2 species:

–Ni–O–Ni–O⋯H–NH_2 H-bonding
–Ni–O–Ni– + OH(a) + NH_2(a)   H-abstraction

It was not possible from the images to decide on the eventual fate of the OH species. One possible pathway also not considered is the recombination of NH_2 radicals and the desorption of hydrazine. If this did occur, then any arguments based on stoichiometries (e.g. OH to NH_2) would not be valid.

Simultaneously with the STM studies, Kulkarni et al.[14] in Cardiff studied by XPS and HREELS the interaction of ammonia with Ni(110)–O and Ni(100)–O surfaces. There was evidence in the N(1s) spectra for more than one nitrogen state present including N(a), but differentiating between NH(a) and NH_2(a) was not possible. The intensity in the N(1s) spectrum region was broad over the range 397–400 eV. As the oxygen coverage increased to >0.3, the oxide O^{2-} component became more prominent and the activity for ammonia oxidation decreased, as was observed by STM. Similar conclusions were reached for water interaction with the Ni(110)–O system.[15]

5.2.5 Ag(110) Pre-exposed to Oxygen

Chemisorption of oxygen at Ag(110) at 300 K forms added rows of –Ag–O– extending along the [001] direction much like those observed with Cu(110). At "saturation" the monolayer, as with Cu(110), has a (2 × 1)O structure.[16] On exposure to ammonia at 300 K, Guo and Madix established[17] that this "oxide" structure undergoes extensive restructuring where the added silver atoms in the monolayer are released to form nanoscale islands with the formation of immobile chemisorbed imide species NH(a). The chemistry of this was established by separate HREELS studies.[18] In an STM movie, the authors showed that the NH species, evident as "mottled" patches, grow initially at the boundaries of the (2 × 1)O islands; simultaneously, silver islands of monatomic height, covered by NH species, are seen to nucleate. The silver islands are fairly randomly distributed over the entire surface, indicating that restructuring of the surface is widespread. There is a net loss of silver atoms during the reaction of ammonia with the (2 × 1)O structure; the silver atoms either nucleate to form islands or migrate to step-edges.[17]

The incorporation of metal atoms into surface chemisorbed structures is, therefore, more widespread than might have been expected from that first observed with oxygen at Cu(110). It is an example of the mobility and incorporation of silver atoms by a molecular fragment NH(a); other examples from the Stanford group are NO_3(a) and SO_3(a) at Ag(110) and SO_3(a) at Cu(110).[19,20]

5.3 Oxidation of Carbon Monoxide

That carbon monoxide could be oxidised in a facile reaction at cryogenic temperature (100 K) was first established in 1987 by XPS at an aluminium surface.[21] The participation of reactive oxygen transients $O^{\delta-}$(s) was central to the mechanism proposed, whereas the chemisorbed oxide O^{2-} state present at 295 K was unreactive. This provided a further impetus for the transient concept that was suggested for the mechanism of the oxidation of ammonia at a magnesium surface (see Chapter 2). Of particular relevance, and of crucial significance, was Ertl's observation by STM in 1992 that oxygen chemisorption at Al(111) resulted in kinetically "hot" adatoms (Figures 4.1 and 4.7).

In 1995, Iwasawa and his colleagues in Tokyo observed similar oxidation chemistry at Cu(110) at low temperatures using a combination of LEED and HREELS.[22] They chose to investigate the reactivity of oxygen states present at the unreconstructed Cu(110) surface between 150 and 200 K, *i.e.* a precursor of the added row (2 × 1)O state. The catalytic formation of CO_2 was facile, occurring with an activation energy of 34.8 kJ mol^{-1}; the active oxygens are suggested to be those not involved in the –Cu–O–Cu– chains. As these chains develop, the rate of CO_2 formation decreased and the authors emphasised the close similarity between the mechanism and that proposed for the oxidation of ammonia at Cu(110). In 1994, Crew and Madix[23] reported an STM study of CO oxidation at a Cu(110) surface precovered with oxygen at 400 K. They concluded that the reaction, which is slow, requiring high exposures (105 L) of CO, appears to occur wholly at the periphery of oxygen islands creating defect sites. Once created, the defects are more reactive and play the dominant role in sustaining the reaction. They later reported[24] STM data for CO oxidation at Cu(110) over the temperature range 150–300 K. Oxygen preadsorbed at 150 K formed Cu–O pseudomolecules which are short, 2–7 units long, and which contrast with the long (2 × 1) chain or island structures characteristic of chemisorbed oxygen at 300 K, both of which are unreactive to CO at 150 K. However, when the surface, with CO *preadsorbed*, is exposed to oxygen at 150 K there is rapid oxidation with desorption of CO_2. Similar oxidation activity had been reported using XPS for adsorbed NH_3 exposed to oxygen at low temperatures,[25] when the preadsorbed chemisorbed oxygen state present at Mg(001) was inactive in ammonia oxidation.

Crew and Madix[24] suggested that a mobile form of oxygen (a transient) is responsible for the oxidation reaction; neither the $(2 \times 1)O$ structure nor what they refer to as "pseudomolecules" are active at low temperature. They also drew attention to the similarity with the model proposed for ammonia oxidation at Cu(110). In 1997, Burghaus and Conrad,[26] using kinetic methods, also suggested that CO oxidation at Ag(110) at low temperatures (100–200 K) was controlled by a highly reactive oxygen state, metastable at 100 K but becoming passive at 200 K, with activity in oxidation only observed above 300 K. This was further evidence for the transient reactive oxygen state.

Guo and Madix[17] discussed in 2003 the implications from STM images observed in real time of the oxidation of CO at Cu(110). Under steady-state conditions at 400 K, both oxygen and carbon monoxide are present at a total pressure of "about 2×10^{-4} Torr" (the proportion of each in the mixture was not stated). There is evidence for reaction anisotropy with CO reacting with oxygen primarily along the rows in the [100] direction, with oxygen states at the end of the rows being much more reactive than within the rows; Monte Carlo simulations suggest a difference in reactivity of a factor of at least 500 between these oxygens.

Ertl and his colleagues in 1997 reported detailed STM data for the oxidation of CO at Pt(111) surfaces, with quantitative rates extracted from the atomically resolved surface events.[27] The aim was to relate these to established macroscopic kinetic data, particularly since it had been shown that no surface reconstruction occurred and the reaction was considered to obey the Langmuir–Hinshelwood mechanism, where it is assumed that the product (CO_2) is formed by reaction between the two adsorbed reactants, in this case O(a) and CO(a). Nevertheless, it was well known that for many features of the CO oxidation reaction at Pt(111) there is no mechanism that is consistent with all features of the kinetics; the inherent problem is that in general a reaction mechanism cannot be uniquely established from kinetics because of the possible contribution of intermediates or complications for which there might be no direct experimental evidence.

A sequence of images (Figure 5.7) were observed at 247 K during the reaction of preadsorbed oxygen adatoms with CO(g). The oxygen adatoms were adsorbed at 96 K then annealed at 298 K, cooled to 247 K and exposed to CO(g) at a pressure of 5×10^{-8} mbar. At this pressure, the CO impact rate corresponds to a monolayer per 100 s and images were observed after 90, 140, 290, 600, 700, 1100 and 2020 s. The initial (at time zero) oxygen structure is (2×2); adsorption of CO resulted first in an increase in the ordering within the (2×2) structure ($t = 140$ s) due to repulsive interactions, with adsorbed CO, presumably due to their mobility. After 290 s and more clearly after 600 s, an

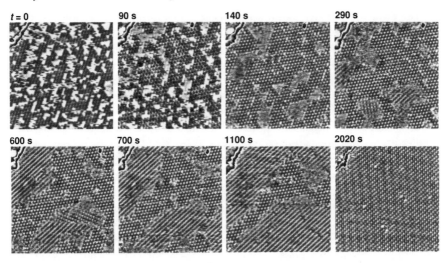

Figure 5.7 STM images (180 × 170 Å) taken during the exposure of a Pt(111)–O adlayer (oxygen exposure 3 L) to carbon monoxide at 247 K and a pressure of 5×10^{-8} Torr. (Reproduced from Ref. 27).

additional ordered streaky structure became visible. This is attributed to the immobile adsorbed CO present in a c(4 × 2) structure, the orientations of which can change with time. With continued exposure to CO the c(4 × 2) areas became more dominant and the number of (2 × 2) oxygen structures decreased as the reaction proceeded, with CO_2 desorbing. Clearly, the two reactants were not randomly distributed but were in separate well-ordered domains with reaction confined to the boundaries between the (2 × 2)O and c(4 × 2)CO domains.

Rates of the reaction based on the distribution of the reactants in the separate domains were determined and shown to be in good agreement with data obtained from classical macroscopic measurements. An activation energy of 11 kJ mol^{-1} was estimated from Arrhenius plots, in good agreement with values from molecular beam studies ($\sim$12 kJ mol^{-1}) and temperature programmed desorption studies (14 kJ mol^{-1}). These STM studies are significant as they represent the first quantitative verification of the macroscopic kinetics of a catalytic reaction by experimentally determined atom resolved studies.

Oxidation of carbon monoxide at a ruthenium surface is an interesting example of where oxidation results in the formation of RuO_2, which provides the platform for the catalytic reaction. The RuO_2(110) surface consists of [001] oriented rows of oxygen atoms, "O-bridge", residing in ruthenium bridge sites in between rows of coordinatively unsaturated Ru sites (Ru-cus). The Ru-cus atoms at the surface have associated with them "dangling bonds", which confers on them specific characteristics. The model described by Kim and Wintterlin[28] for CO oxidation at RuO_2(110) is as follows. A CO molecule bonds to the Ru-cus atom to form an unstable (reactive) "CO-cus" state, which reacts with a neighbouring O-bridge atom to form CO_2, which desorbs. The

O-bridge vacancy created can than be occupied by a further CO molecule from the gas phase to form a CO-bridge. Further O-cus atoms can react with the neighbouring CO-bridge to form $CO_2(g)$, the vacancies in the O-bridge sites being occupied by further O atoms (from the gas phase), which results in the original structure of the stoichiometric RuO_2 overlayer being re-formed. Two reaction pathways for CO oxidation therefore exist, CO-cus reacting with O-bridge atoms and CO-cus reacting with O-cus. An alternative way of describing O-cus is that it is a precursor state to the fully oxidised state O^{2-} and with a smaller negative change and assigned as $O^{\delta-}$. In this sense, it can be regarded as a "transient oxygen state" and analogous to that exhibiting "high oxidation activity" in catalytic oxidation reactions at *metal* surfaces.

Two different STM experiments were carried out by Kim and Wintterlin:[28] CO(g) reacting with stoichiometric $RuO_2(110)$ and $O_2(g)$ reacting with CO-cus preadsorbed at the surface. The STM images (Figure 5.8) show that both of these reactions – O-bridge with O-cus and O-cus with CO-cus – are essentially random processes. Only during the early stages of the latter reaction, when the surface is saturated with O-cus, do vacancies have a role; in all other respects,

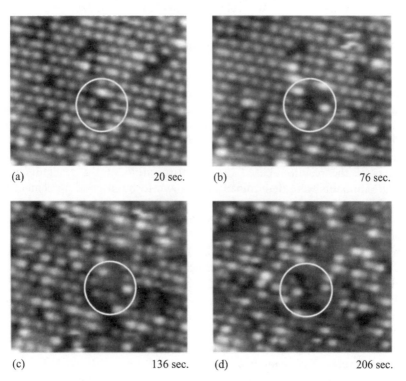

Figure 5.8 Images observed during the adsorption of O_2 on the CO-saturated $RuO_2(110)$ surface. Bright dots are CO-cus molecules along the <001> direction; dark sites in (a) are vacancies. The circle shows the development of a vacancy with time at an oxygen pressure of 2×10^{-8} Torr at room temperature. (Reproduced from Ref. 28).

$RuO_2(110)$ exemplifies Langmuirian behaviour where the catalyst surface consists of equivalent sites statistically occupied by the reactants. This contrasts markedly with catalytic oxidation at metal surfaces, where oxygen transients, high surface mobility and island structures are dominant. The difference is in the main attributed to differences in surface diffusion barriers at metal and oxide surfaces.

5.4 Oxidation of Hydrogen

There have been extensive spectroscopic (XPS, UPS and HREELS) studies of oxygen–water interactions at metal surfaces involving both preadsorbed oxygen and coadsorption of oxygen–water mixtures. In general, the conclusions can be summarised as follows: (a) a complete oxygen adlayer is unreactive to water vapour at room temperature and below; (b) the partially complete oxygen adlayer shows some limited reactivity in hydrogen abstraction and hydroxyl formation; and (c) with water–oxygen mixtures (coadsorbed) facile hydroxyl formation is observed at low temperatures resulting in a fully hydroxylated surface. Some of the earlier data were discussed in 1983 by Carley et al.[29] (see also Figure 2.2).

Volkening et al.[30] studied by STM hydrogen oxidation at a Pt(111)–O surface, the oxygen being adsorbed under a constant hydrogen pressure at low temperatures (110–300 K). A series of images were recorded below 170 K; terraces were covered with bright spots and a bright ring whose circumference expanded between two images taken 625 s apart. The ring appeared to travel with constant velocity and without changing its shape, reminiscent of reaction fronts in non-linear diffusion reactions.

The interaction of hydrogen with preadsorbed oxygen at Pt(111) led to hexagonal and honeycomb structures to develop at 131 K, which could be associated with OH phases with also evidence for water formation. The front (bright ring) consisted mainly of OH(a) and the area behind the front of $H_2O(a)$. The mechanism suggested is that H(a) reacts first with O(a) to form OH(a) and then $H_2O(a)$; the water is mobile and reacts with O(a) to form OH(a); it is therefore an autocatalytic reaction.

$$O(a) + H(a) \rightarrow OH(a)$$

$$OH(a) + H(a) \rightarrow H_2O(a)$$

$$H_2O(a) + O(a) \rightarrow 2OH(a)$$

At "high" temperatures" (>170 K), the water desorbs and so the autocatalytic reaction cannot be sustained and is an explanation for why the $H_2 + O_2$ reaction slows, the formation of OH species now being solely dependent on the H(a) + O(a) reaction, which is the slowest step in the above scheme. That the water + oxygen reaction was "fast" and facile was evident from the spectroscopic studies at both nickel and zinc surfaces, when the oxygen surface coverage was low and involving isolated oxygen adatoms.

Studies of the interaction of hydrogen with "oxidised" nickel surfaces have long contributed to the understanding of chemisorption and catalysis, the presence of surface oxygen being associated with "slow" or activated adsorption (for example) by Schuit and de Boer[31] in 1951 and Morrison[32] in 1955. Furthermore, a pressure dependence of $p^{1/2}$ indicated that the reactive hydrogen is the dissociated state. In the absence of surface oxygen, hydrogen dissociation is non-activated and "fast" at a nickel surface at 80 K. What, then, have we learnt from STM at the atom resolved level?

At the University of Aarhus, Sprunger et al.[33] studied in detail the reaction of hydrogen with preadsorbed oxygen at Ni(110) at 300 and 470 K at various oxygen coverages. The Ni(110)–oxygen system had been previously studied by the authors with STM evidence for (3×1) and (2×1) structures comprising –Ni–O– added rows running along the [001] direction; at low oxygen coverage, the –Ni–O– chains are relatively short and mobile at room temperature. At higher oxygen exposure and a surface coverage approaching a monolayer, a (9×5) structure followed by epitaxial NiO(100) is observed. For oxygen coverages less than 0.5, hydrogen exposure at 300 K results in what the authors describe as "streaky" (1×2) –H structures developing along the $<110>$ direction. The nucleation sites of this hydrogen-induced phase are homogeneously spread over large terraces. Simultaneously with this there occurs a compression of the oxygen structures into (2×1)O domains.

The growth of the hydrogen-induced phase is anisotropic with islands of no more than 1–2 nm along the $<001>$ direction but much longer along the close-packed direction. The authors propose a complex mechanism for the growth of the H-induced reconstructed phase; as hydrogen dissociatively adsorbs along the chains, additional nickel atoms are "pulled out" of the terraces or step-edges, diffuse along the missing and/or added rows and then are incorporated at the end of the growing –Ni–H– chains. These appear as "bright stripes" in the STM image; the "dark stripes" are "missing row" –Ni–H– structures (Figure 5.9). Sequential images indicate that as the –Ni–H– rows develop the diffusing –Ni–O– rows become locally compressed into (2×1)O structures. We have, therefore, an example of island segregation, evidence that would be difficult to extract from what would be a streaky diffuse pattern observed by LEED. Oxygen coverages remain unchanged after exposure to $H_2(g)$ at 300 K, indicating that a "titration" reaction does not occur. When the oxygen pre-coverage is at or above 0.5 monolayers, no change is observed in the (2×1), (3×1) and (9×5) structures when exposed to hydrogen at 300 K, a characteristic feature of the inactivity of the "complete" oxide monolayer at nickel surfaces at room temperature. For oxygen coverages less than 0.66 monolayers, total titration of the oxygen (removal as desorbed water) occurs at 470 K, the rate being 10^3 times faster than at 295 K. Reaction is initiated at large (2×1) terraces, width $\geq$ 50nm, with dark holes observed which are 2–5 units of –Ni–O– in length. After a hydrogen exposure of 190 L, all the oxygen is removed and the added nickel atoms are incorporated to the surface at step-edges.

Two reaction regimes are recognised, an "induction period", when little is observed to take place, followed by a "reaction period", which is rapid. The

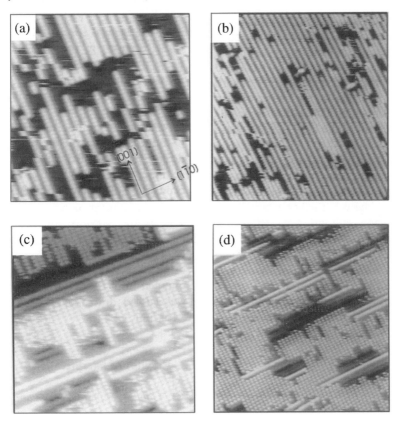

Figure 5.9 Images of the Ni(110)–O surface for two oxygen coverages before and after exposure to H_2 at 300 K. (a) $\theta = 0.2$ ML and (c) after H_2 exposure; (b) $\theta = 0.33$ ML and (d) after H_2 exposure. (Reproduced from Ref. 29).

extent in time, for a given hydrogen pressure, of the induction period is greater for larger oxygen coverages. The authors concluded from studies of surfaces of different morphology – flat terraces or high density of steps – that the latter greatly enhances the local titration rate. The terminating ends of the added –Ni–O– rows are proposed as the "titration active" sites similar to that observed at the Cu(110)–O surface where, on exposure to ammonia, the termination of the –Cu–O– chains at step-edges was "decorated" by NH species with water desorbed.

5.5 Oxidation of Hydrocarbons

In 1978, Wachs and Madix[34] drew attention to the role of oxygen in the oxidation of methanol being "not completely understood" at copper surfaces. They established the role of methoxy species as the favoured route to the formation of formaldehyde and that "to a lesser extent some methanol was oxidised to formate, which subsequently decomposed to CO_2 and H_2". This was an elegant kinetic study relying on deuterated species to provide evidence on the reaction pathways. In 1996, Carley et al.[35] established that the formation of formate could, depending on the experimental conditions, be the major pathway at Cu(110), and in 1997, using the coadsorption approach, Davies and Mariotti[36] showed that the selectivity to formaldehyde or formate could be controlled by varying the methanol to oxygen ratio in the gas mixture. With a methanol-to-oxygen ration of 5:1, only formate is produced, whereas with a dioxygen-rich (1:1) mixture formaldehyde is the major product. The authors make the significant comment that on the basis of kinetic and spectroscopic (XPS) studies it is "the microscopic structure of the surface" under reaction conditions – islands of methoxy and oxygen states – that control the selectivity. This model was further established by Poulston et al.[37] through STM studies. In a paper submitted later than that by Davies and Mariotti, these authors established (Figure 5.10) the presence of surface structures associated with methoxy (5×2), formate (3×1) and $c(2 \times 2)$ and the oxide reconstructed state (2×1). Control of reaction pathways is, therefore, intimately linked to the nanoscale island structures, that is to the relative concentrations and surface distribution of the methoxy, the active: (isolated) oxygen states and the comparatively unreactive oxide, $(2 \times 1)O$ islands. The similarities with other surface-catalysed oxidation reactions, especially that of ammonia, is striking, with the following being the steps evident from both STM and kinetic studies:

$$2CH_3OH(g) + O^{\delta-}(a) \rightarrow 2CH_3O(a) + H_2O(g)$$

$$CH_3O(a) \rightarrow H_2CO(a) + H(a)$$

$$H_2CO(a) \rightarrow H_2CO(g)$$

$$H_2CO(a) + O^{\delta-}(a) \rightarrow HCO_2(a) + H(a)$$

$$O^{\delta-}(a) \rightarrow \text{"oxide"}$$

$$2H(a) \rightarrow H_2(g)$$

Although it was not possible to distinguish between the specificity of oxygen states responsible for methoxy and formate formation, they were to be associated with isolated oxygen adatoms and oxygen states present at the periphery of the $(2 \times 1)O$ islands.

Spectroscopic studies of propene oxidation at an Mg(0001) surface at 295 K in 1987 indicated that $O^{\delta-}(s)$ transient oxygen states, present only during the early stage of the reaction, were active in the formation of C_4, C_6 and C_7 gaseous products. The surface becomes inactive as reaction progresses, with evidence from the Mg(1s), C(1s) and O(1s) photoelectron spectra for the development of Mg^{2+} and carbonaceous species, including carbonate.[38] An STM investigation coupled with *in situ* XPS established (Figure 5.11) that at 295 K, oxidation activity leading to the formation of the gaseous hydrocarbons

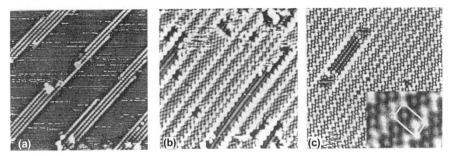

Figure 5.10 STM images of a Cu(110)–O surface (a), after exposure (10 L) to CH$_3$OH at 270 K (b) and 40 min later (c). Note the transformations of the (2 × 1)O strings into zig-zag chains and c(2 × 2) structures (b) and with time the oxygen has been removed and the surface evolved into a (5 × 2) methoxy reconstruction (c). (Reproduced from Ref. 37).

was confined to the early stage of oxide nucleation.[39] During this stage, $O^{\delta-}$ species formed by the dissociative chemisorption of oxygen are highly mobile and reactive, whereas the O^{2-} state associated with the oxide hexagonal structures and characterised by the shift in the Mg(1s) binding energy, reflecting the presence of Mg^{2+}, is inactive in H-abstraction and with its development the reaction is poisoned. Gaseous oxidation products are, therefore, only observed during that state of surface oxidation where $O^{\delta-}$ states predominate with a radical-type reaction involved resulting in C–C bond cleavage, H-abstraction and dimerisation:

$$O^{\delta-}(s) + C_3H_6(g) \rightarrow CH_2=CH-CH_2(s) + OH(a) \quad \text{H-abstraction}$$
$$CH_2=CH-CH_2(s) \xrightarrow{O^{\delta-}} C_6H_6(g) \quad \text{dimerisation}$$
$$CH_2=CH-CH=CH_2 \quad \text{coupling reaction}$$
$$CO_3 \quad \text{complete oxidation}$$

Which reaction pathway dominates is dependent on the $O^{\delta-}$(s) concentration and the relative rates of carbonate formation and desorption of the gaseous C_4, C_6 and C_7 gaseous products; some control of these is possible[38] by varying the propene-to-oxygen ratio and also the oxidant, such as substituting N_2O for O_2.

An interesting and significant example of where a metal atom is incorporated into an inorganic molecular intermediate is that of the formation of acetylide (C_2) on Ag(110). The acetylide is formed by exposing an Ag(110)–(2 × 1)O overlayer to acetylene; the reaction is facile, with a reaction probability of unity at room temperature. The first STM study was by Guo and Madix[40] in 2004, but the Stanford group had investigated the reaction earlier by LEED, XPS and NEXAFS. It was a further example of a facile oxydehydrogenation reaction under conditions where the clean metal was unreactive:

$$C_2H_2 + O(a) \rightarrow C_2(a) + H_2O(g)$$

The presence of surface acetylide species was confirmed by titration with acetic acid, where the C_2 species are removed as acetylene and replaced by

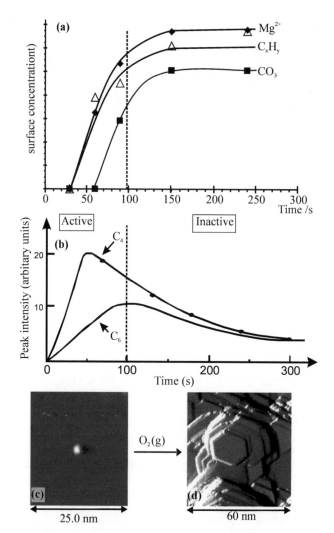

Figure 5.11 Variation in the catalytic activity of an Mg(0001) surface when exposed to a propene-rich propene–oxygen mixture at room temperature. The surface chemistry is followed by XPS (a), the gas phase by mass spectrometry (b) and surface structural changes by STM (c, d). Initially the surface is catalytically active producing a mixture of C_4 and C_6 products, but as the surface concentrations of carbonate and carbonaceous C_xH_y species increase, the activity decreases. STM images indicate that activity is high during the nucleation of the surface phase when oxygen transients dominate. (Reproduced from Ref. 39).

chemisorbed acetate:

$$C_2(a) + 2CH_3COOH \rightarrow C_2H_2(g) + 2CH_3COO(a)$$

The STM study provided microscopic evidence for the participation of added silver atoms in the chemisorbed acetylide structure. The Ag(110)–p(2 × 1)O layer was formed by exposing a clean Ag(110) surface to oxygen at 300 K; the (2 × 1)O overlayer occupied approximately 89% of the surface with 11% remaining "clean". A real-time movie of the reaction of this surface with acetylene was taken (Figure 5.12). After 10 s of exposure at a pressure of 2×10^{-9} Torr, large protrusions (either C_2 or C_2H_2 species) appear on top of the oxygen (2 × 1) domains. With increasing time (exposure), the (2 × 1) areas decrease, the oxygen adatoms being removed as water, and new protrusions appear in the boundary regions. These protrusions develop into rows along the $\langle 1\bar{1}0 \rangle$ axis. After a total exposure of 2 L, all the oxygen rows have been replaced by rows, some thicker than others. The acetylides develop as row structures along the $\langle 1\bar{1}0 \rangle$ axis, which with increasing coverage are compressed along the <100> axis to form what Guo and Madix describe as "normal" p(2 × 2), p(2 × 3), p(2 × 1) and p(14 × 1) structures. Some of these structures are transition phases, the p(2 × 3) a transition between p(2 × 3) and p(2 × 1). The final "surface-saturated" acetylide structure incorporates about 0.5 ML of silver atoms – confirmed by titration with acetic acid.

A line profile analysis of the saturated acetylide surface reveals the buckled nature of the overlayer with a periodicity of seven protrusions or "14 lattice units" along the $<1\bar{1}0>$ axis. The nominal "14" units actually match 13 lattice units; therefore to accommodate seven protrusions on 13 lattice units with equal spacing would result in surface buckling (Figure 5.13). The distance between two terminal silver atoms is 5.37 Å, which is 2% shorter than that in silver acetylide based on the assumption of covalent radii.

At an Ni(110)–O surface exhibiting a (3 × 1) structure (0.3 ML of oxygen), benzene adsorption at room temperature induces a compression of the (3 × 1) added-row to a (2 × 1) structure. There was no evidence for a direct reaction between the surface oxygen and benzene, but on heating to 600 K the oxygen is removed (as CO) and the surface is clean, other than areas of a p(4 × 5)C carbidic phase.[41]

5.6 Oxidation of Hydrogen Sulfide and Sulfur Dioxide

That chemisorbed oxygen was active in hydrogen abstraction, resulting in water desorption and the formation of chemisorbed sulfur, was first established by XPS at copper and lead surfaces.[42] An STM study of the structural changes when a Cu(110)–O adlayer is exposed (30 L) to hydrogen sulfide at 290 K indicates the formation of c(2 × 2)S strings.

The Aarhus group observed similar structural changes when Ni(110)–O was exposed to H_2S. However, the c(2 × 2)S structure that forms transforms at high exposures into a more stable (4 × 1)S adlayer. The removal of oxygen from the (2 × 1)O state as desorbed water results in the development of low-coordinated rows of nickel atoms which coalesce to form Ni(1 × 1) islands leaving behind Ni(1 × 1) troughs. Sulfur chemisorbs on these islands with a c(2 × 2) structure. The islands and troughs have an apparent height and depth relative to the (2 × 1)O terrace of 0.75 and 0.5 Å, respectively.

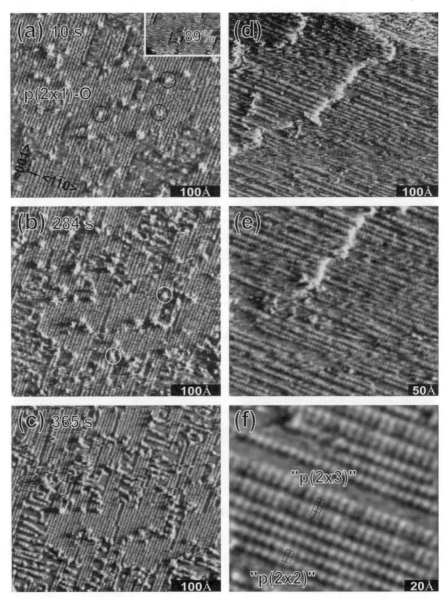

Figure 5.12 STM images during acetylene reaction with Ag(110)–p(2 × 1)O at 300 K. (a) The surface 10 s after exposure to C_2H_2 at a pressure of 2×10^{-9} Torr; (b) the same region 274 s later; (c) the same region 81 s later; (d)–(f) progressively magnified images of the surface after an exposure of 2 L. Acetylide forms thin p(2 × 2) and thick p(2 × 3) rows running at right-angles to the –Ag–O– rows. (Reproduced from Ref. 40).

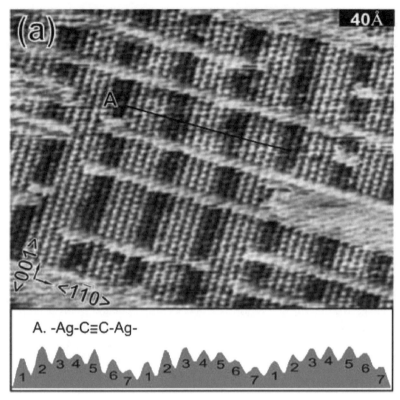

Figure 5.13 High-resolution image of a surface-saturated acetylide overlayer at 300 K. Below the image is a cross-section along the line A. Note the buckling of the adlayer. (Reproduced from Ref. 40).

Whether the motion of adsorbates (in general) is fragmented, a collective motion of an entire row or as a series of correlated jumps of individual atoms has been debated for some years. In a study of the oxidation of sulfur dioxide at a Cu(110)–O surface ($\theta_0 = 0.24$ ML) at 300 K Alemozafar et al.[43] concluded from an STM movie that surface mobility involved the collective motion of an entire row, the –Cu–O–Cu– rows in the <100> direction and the SO_3 rows in the <110> direction. The Stanford group, using a combination of horizontally and vertically scanned images, provided the crucial evidence for the collective motion of an entire –Cu–O–Cu– string. Single fragmented or individual motion would have resulted in zig-zag structures being observed in the STM image along the <110> direction, irrespective of the scanning direction. No such images were observed. Similarly, images of SO_3 rows also indicated their collective motion. The authors draw attention to the many examples of collective motion in both surface reactions and nature due to "weak" interaction energies.

At Ag(110), sulfur dioxide reacts with a p(2 × 1)O oxygen adlayer to give a c(6 × 2) sulfite structure with six sulfite (SO_3) species associated with each unit cell. The authors suggest that there is a rearrangement of the "added" silver in the (2 × 1)O structure to precipitate new islands of sulfite. On heating to 600 K the surface is transformed to sulfate with one sulfate species occupying each of the (3 × 2) unit cells according to the following stoichiometry:

$$6SO_3(a) \rightarrow 4SO_2(g) + 2SO_4(a) + 2O(inc)$$

Since no evidence was obtained for any *surface* oxygen, it is assumed to be incorporated, O(inc), and subsurface.

5.7 Theoretical Analysis of Activation by Oxygen

The experimental evidence, first based on spectroscopic studies of coadsorption and later by STM, indicated that there was a good case to be made for transient oxygen states being able to open up a non-activated route for the oxidation of ammonia at Cu(110) and Cu(111) surfaces. The theory group at the Technische Universiteit Eindhoven considered[5] the energies associated with various elementary steps in ammonia oxidation using density functional calculations with a Cu(8,3) cluster as a computational model of the Cu(111) surface. At a Cu(111) surface, the barrier for activation is $+344\,kJ\,mol^{-1}$, which is insurmountable; copper has a nearly full d-band, which makes it difficult for it to accept electrons or to carry out N–H activation. Four steps were considered as possible pathways for the initial activation (dissociation) of ammonia (Table 5.1).

In the presence of adsorbed atomic oxygen, the activation barrier is substantially lowered to $132\,kJ\,mol^{-1}$, the reaction is endothermic at $48\,kJ\,mol^{-1}$, but the high activation energy suggests that the N–H bond would not be broken. However, at high temperatures it might be achieved.

A molecular oxygen state is the most likely to be involved, it would require a barrier of only $67\,kJ\,mol^{-1}$ and is exothermic; a hydroperoxide state is formed together with $NH_2(a)$. When the heats of adsorption of ammonia and oxygen are taken account of, then according to Neurock[44,45] there is no apparent activation barrier to N–H activation.

Similar calculations established that oxygen-assisted acetate decomposition was consistent with the experimental work of Davis and Barteau.[46]

The density function calculations for the ammonia oxidation reaction do, however, depend on models where the reactants are in stable adsorption states

Table 5.1 Possible pathways for initial activation of ammonia.

Reaction	Activation energy $(kJ\,mol^{-1})$	Overall reaction energy $(kJ\,mol^{-1})$
$NH_3(g) \rightarrow NH_2(g) + H(g)$	+498	+498
$NH_3^* \rightarrow NH_2^* + H^*$	+344	+176
$NH_3^* + O^* \rightarrow NH_2^* + OH^*$	+132	+48
$NH_3^* + O^* \rightarrow NH^* + H_2O(g)$	>200	+92
$NH_3^* + O_2^* \rightarrow NH^*O^* + H_2O(g)$	+67	−84
$NH_3^* + O_2^* \rightarrow NH^* + O^* + H_2O(g)$	+134	−184

and not undergoing rapid surface diffusion, not thermally accommodated and mimicking a two-dimensional gas reaction. It was the last model that was the one proposed on the basis of the spectroscopic coadsorption studies (Chapter 2) and supported later by STM. In this case, it is not clear what energy parameters should be used for the adsorbates, and this may be a general problem in applying theory to heterogeneously catalysed reactions where surface mobility and surface disorder appear to play a significant role in the catalytic reaction. STM has played an important role in highlighting this.

5.8 Summary

Oxygen activation of molecules at metal surfaces was first established in the 1970s by surface spectroscopies (XPS and UPS) over a wide temperature range (80–400 K). Furthermore, the distinction was made between the reactivity of partially covered surfaces and the relative inactivity of the "oxide" monolayer.

Although the significance of oxygen transient states in being able to influence catalytic oxidation pathways at *metal* surfaces was well established by spectroscopic methods, there is also evidence that at *oxide* surfaces coadsorbing dioxygen with adsorbates (HX) such as ethane, propene and ammonia, can induce peroxide ions. Giamello et al.[49] in 1989 established by ESR the formation of superoxide ions at MgO. The O_2^- ion being formed by electron transfer from the radical anion to molecular oxygen:

$$XH + O^{2-}(\text{surf}) \rightarrow X^- + OH^- \quad \text{hydrolytic adsorption}$$
$$X^- + O_2 \rightarrow X + O_2^- \quad \text{electron transfer}$$

In view of the spectroscopic evidence available, particularly from coadsorption studies (see Chapter 2), ammonia oxidation at Cu(110) became the most thoroughly studied catalytic oxidation reaction by STM. However, a feature of the early STM studies was the absence of *in situ* chemical information. This was a serious limitation in the development of STM for the study of the *chemistry* of surface reactions. What, then, have we learnt regarding oxygen transient states providing low-energy pathways in oxidation catalysis?

In the early 1990s, groups at Aarhus University and the Fritz Haber Institute established for metal–oxygen systems that facile mobility of both the metal substrate atoms and the oxygen adsorbate was a feature of the chemisorption process. Different oxygen states at metal surfaces were characterised with STM and *in situ* XPS by their different reactivities. In the case of Cu(110), disordered oxygen metastable states present at 100 K were active in the complete stripping of hydrogen from ammonia but became inactive as the temperature was raised, when they underwent a disorder–order transition to form the stable $(2 \times 1)O$ reconstructed overlayer. Similarly, propene oxidation to give C_4 and C_6 gaseous products at Mg(0001) at room temperature is confined to the oxygen transient and mobile states present during the oxide nucleation stage. There is now much evidence that this may well be a general phenomenon in the reactivity of oxygen states at metal surfaces and could provide a key to the understanding of catalytic oxidation chemistry with, for example, nascent oxygen states, $O^{\delta-}$ (s), invoked in the oxidation of methane.[47,48] Correlations between the atom resolved structures with chemical reactivity of the oxygen states was a significant step forward in the development of nanosurface chemistry. Collective motion of adsorbates was widely debated, with Guo and Madix establishing that it occurred in a number of surface reactions including sulfite formation at Cu(110)–O, with metal incorporation into intermediate states observed for both ammonia and acetylene oxidation at Ag(110)–O. Although questions had been raised[25] in 1987 as to whether Eley–Rideal and Langmuir–Hinshelwood mechanisms were appropriate models for surface reactions involving mobile "hot" oxygen transients, STM provided conclusive evidence for their limitations.

The ultimate experiment in STM studies of surface reactions is to achieve real-time videos; this has rarely been achieved. Ertl and Madix are the exceptions, with their investigations of CO oxidation at Pt(111) and NH_3 oxidation at Cu(110). Wintterlin *et al.* emphasise the limitation of deriving kinetic equations to describe catalytic oxidation reactions based on macroscopic parameters (pressure, temperature and concentrations). Nevertheless, for CO oxidation at Pt(111), the kinetic parameters derived from classical macroscopic rates are almost identical with those based on kinetic models derived from atomic-scale data. It is, however, well recognised that kinetics alone cannot unambiguously establish a reaction mechanism, in spite of their usefulness, particularly in the development of early views on catalytic mechanisms and in chemical reactor design. What is also true is that to build detailed models based on STM images alone can be speculative, an image being a convolution of local atomic and electronic structure.

References

1. M. W. Roberts, *Adv. Catal.* 1980, **29**, 55, and references cited therein.
2. W. M. H. Sachtler, in *Surface Chemistry and Catalysis*, ed. A. F. Carley, P. R. Davies, G. J. Hutchings and M. S. Spencer, Kluwer Academic/Plenum Press, New York, 2002, 207.
3. Reviewed in M. W. Roberts, *Surf. Sci.*, 1994, **299/300**, 769; *Chem. Soc. Rev.*, 1996, 437.
4. A. F. Carley, P. R. Davies and M. W. Roberts, *Catal. Lett.*, 2002, **80**, 25.
5. M. Neurock, R. A. Van Santen, W. Biemolt and A. P. Jansen, *J. Am. Chem. Soc.*, 1994, **116**, 6860.
6. B. Afsin, P. R. Davies, A. Pashusky, M. W. Roberts and D. Vincent, *Surf. Sci.*, 1993, **284**, 109; A. Boronin, A. Pashusky and M. W. Roberts, *Catal. Lett.*, 1992, **16**, 345.
7. A. F. Carley, P. R. Davies, M. W. Roberts and D. Vincent, *Top. Catal.*, 1994, **1**, 35.
8. A. F. Carley, P. R. Davies, K. R. Harikumar, R. V. Jones, G. U. Kulkarni and M. W. Roberts, *Top. Catal.*, 2001, **14**, 101.

9. X.-C. Guo and R. J. Madix, *Faraday Discuss.*, 1996, **105**, 139.
10. X.-C. Guo and R. J. Madix, *Surf. Sci. Lett.*, 1996, **367**, L95.
11. A. F. Carley, P. R. Davies and M. W. Roberts, *Chem. Commun.*, 1998, 1793.
12. A. F. Carley, P. R. Davies, K. R. Harikumar, R. V. Jones and M. W. Roberts, *Top Catal.*, 2003, **24**, 51.
13. L. Ruan, I. Stensgaard, E. Laegsgaard and F. Besenbacher, *Surf. Sci.*, 1994, **314**, L873.
14. G. U. Kulkarni, C. N. R. Rao and M. W. Roberts, *J. Phys. Chem.*, 1995, **99**, 3310.
15. G. U. Kulkarni, C. N. R. Rao and M. W. Roberts, *Langmuir*, 1995, **11**, 2572.
16. J. V. Barth, T. Zambelli, J. Wintterlin and G. Ertl, *Chem. Phys. Lett.*, 1997, **270**, 152.
17. X.-C. Guo and R. J. Madix, *Acc. Chem. Res.*, 2003, **36**, 471.
18. D. M. Thornburg and R. J. Madix, *Surf. Sci.*, 1989, **220**, 268.
19. X.-C. Guo and R. J. Madix, *Surf. Sci.*, 2002, **496**, 39.
20. A. R. Alemozafar, X.-C. Guo, R. J. Madix, N. Hartmann and J. Wang, *Surf. Sci.*, 2002, **504**, 223.
21. A. F. Carley and M. W. Roberts, *J. Chem., Soc. Chem. Commun.*, 1987, 355.
22. T. Sueyoshi, T. Sasaki and Y. Iwasawa, *Chem. Phys. Lett.*, 1995, **241**, 189.
23. W. W. Crew and R. J. Madix, *Surf. Sci. Lett.*, 1994, **314**, L34.
24. W. W. Crew and R. J. Madix, *Surf. Sci.*, 1996, **356**, 1.
25. C. T. Au and M. W. Roberts, *J. Chem. Soc. Faraday Trans. 1*, 1987, **83**, 2047; General Discussion, 2085.
26. U. Burghaus and H. Conrad, *Surf. Sci.*, 1997, **370**, 17.
27. J. Wintterlin, S. Völkening, T. V. W. Janssens, T. Zambelli and G. Ertl, *Science*, 1997, **278**, 1931.
28. S. H. Kim and J. Wintterlin, *J. Phys. Chem. B.*, 2004, **108**, 14565.
29. A. F. Carley, S. Rassias and M. W. Roberts, *Surf. Sci.*, 1983, **135**, 35.
30. S. Volkening, K. Bedürftig, K. Jacobi, J. Wintterlin and G. Ertl, *Phys Rev. Lett.*, 1999, **83**, 2672; see also J. Wintterlin, *Adv. Catal.*, 2000, **45**, 131.
31. G. C. A. Schuit and N. H. de Boer, *Recl. Trav. Chim. Pays-Bas.*, 1951, **70**, 1067.
32. S. R. Morrison, *Adv. Catal.*, 1955, **7**, 50.
33. P. T. Sprunger, Y. Okawa, F. Besenbacher, I. Stensgaard and K. Tanaka, *Surf. Sci.*, 1995, **344**, 98.
34. I. Wachs and R. J. Madix, *J. Catal.*, 1978, **53**, 208.
35. A. F. Carley, P. R. Davies, G. G. Mariotti and S. Read, *Surf. Sci.*, 1996, **364**, L525.
36. P. R. Davies and G. G. Mariotti, *Catal. Lett.*, 1997, **46**, 133.
37. S. Poulston, A. H. Jones, R. A. Bennett and M. Bowker, *J. Phys.: Condens. Matter.*, 1996, **8**, L765.
38. C. T. Au, X. C. Li, J. A. Tang and M. W. Roberts, *J. Catal.*, 1987, **106**, 538.
39. A. F. Carley, P. R. Davies and M. W. Roberts, *Philos. Trans. R. Soc. London Ser. A.*, 2005, **363**, 829.

40. X.-C. Guo and R. J. Madix, *Surf. Sci.*, 2004, **564**, 21.
41. I. Stensgaard, L. Ruan, E. Laegsgaard and F. Besenbacher, *Surf. Sci.*, 1995, **337**, 190.
42. K. Kishi and M. W. Roberts, *J. Chem. Soc., Faraday Trans. 1*, 1975, **71**, 1715; M. W. Roberts, L. Moroney and S. Rassias, *Surf. Sci.*, 1981, **105**, L199.
43. A. R. Alemozafar, X.-C. Guo and R. J. Madix, *J. Chem. Phys.*, 2002, **116**, 4698.
44. M. Neurock, R. A. Van Santen, W. Beimolt and P. A. Jansen, *J. Am. Chem. Soc.*, 1994, **116**, 6860.
45. M. Neurock, in *Dynamics of Surfaces and Reaction Kinetics in Heterogeneous Catalysis*, ed. G. F. Froment and K. C. Waugh, Elsevier, Amsterdam, 1997, 3.
46. J. L. Davies and M. A. Barteau, *Surf. Sci.*, 1991, **256**, 50.
47. S. Kameoka, T. Nobukawa, S. Tanaka, S. Ito, K. Tomishige and K. Kanimori, *Phys. Chem. Chem. Phys.*, 2003, **5**, 3238.
48. G. J. Hutchings, M. S. Scurrell and J. R. Woodhouse, *J. Chem. Soc., Chem. Commun.*, 1987, 1388.
49. E. Giamello, P. Ugliengo and E. Garrone, *J. Chem. Soc., Faraday Trans. 1*, 1989, **85**, 1373.

Further Reading

G. Centi, F. Cavani and F. Trifiró, *Selective Oxidation by Heterogeneous Catalysis*, Kluwer Academic/Plenum Press, New York, 2001.

S. T. Oyama and J. W. Hightower (eds), *Catalytic Selective Oxidation*, ACS Symposium Series, Vol. 523, American Chemical Society, Washington, DC, 1992.

W. Buijs, Challenges in oxidation catalysis, *Top. Catal.*, 2003, **24**, 73.

E. Broclawik, J. Haber, Quantum chemical study of the reaction of ammonia with transient oxygen species, *J. Mol. Catal.*, 1993, **82**, 353.

S. Golunski, R. Rajaram, Catalysis at lower temperatures, *Cattech*, Kluwer Academic/Plenum, New York, 2002, **6**, 30.

R. Mason, Catalysis in chemistry and biochemistry, *Catal. Lett.*, 2004, **98**, 1.

M. Che, A. J. Tench, Characterisation and reactivity of mononuclear oxygen species on oxide surfaces, *Adv. Catal.*, 1982, **31**, 78.

CHAPTER 6
Surface Modification by Alkali Metals

"The secret of science is to ask the right question and it is the choice of problem more than anything that marks the man of genius in the scientific world"

C. P. Snow

6.1 Introduction

Interest in the chemistry and physics associated with alkali metal modification of surfaces stems from a number of technologically significant areas in addition to heterogeneous catalysis, including photoemitting cathodes, battery components and electrochemistry. Although in catalysis the alkali metal is usually added as a "compound", fundamental studies in surface science have been mainly confined to the role of the alkali metal in controlling the physics and chemistry of the modified surface. It is the alkali metal's ability to control reaction pathways, and therefore selectivity in catalysis, that is most significant, well established in ammonia synthesis and Fischer–Tropsch and water-gas shift reactions.[1] With the advent of surface science, alkali metal systems were investigated by surface spectroscopies with carbon monoxide, nitrogen and oxygen adsorbates studied extensively in view of their industrial and fundamental relevance.[2] Nitrogen chemisorption at iron surfaces was considered to be "slow and activated" by classical adsorption methods in the 1960s, with also good experimental evidence that alkali metal oxides such as K_2O acted as a promoter in ammonia synthesis over an iron catalyst. An electronic factor was suggested to operate.[3] Ozaki *et al.*[4] demonstrated that metallic potassium present at iron surfaces increases the activity even more than K_2O. Ertl *et al.*[3] investigated the influence of potassium on the chemisorption of nitrogen on Fe(100) at low pressure and showed that a concentration of $1.4 \times 10^{14}\,cm^{-2}$ of potassium increases the rate by a factor of about 300 at 450 K with negligible activation energy (Figure 6.1). There was no evidence in the LEED pattern for

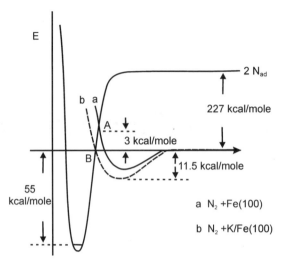

Figure 6.1 Schematic potential energy diagram for atomic and molecular nitrogen adsorption on a clean and K-covered Fe(100) surface. Curve (a) is for N_2 + Fe(100); curve (b) is for N_2 + Fe(100)–K. Note the lowering of the activation energy for dissociation from 3 kcal mol^{-1} to zero. (Reproduced from Ref. 3).

any ordering of the Fe(100)–K adlayer, but the work function in the presence of potassium decreased by 2.25 eV for a potassium concentration of 4×10^{14} cm^{-2}, supporting the concept of the electronic factor in enhancing the rate of nitrogen dissociative chemisorption. The activation by potassium was interpreted as the lowering of the activation energy and applying the Lennard-Jones model for adsorption (Figure 6.1).

A recent example of how reaction pathways can be controlled in selective oxidation by alkali metals is that of propene oxidation at Ag(100)–Cs. Although the Ag(100) surface is unreactive at 290 K, an Ag(100) surface partially covered by caesium is very reactive to a propene–oxygen mixture, resulting in complete oxidation to give surface carbonate at 160 K. For caesium alone dehydrogenation is the main pathway but with some evidence for selective oxidation.[5] In neither of these systems, Fe(100)–K and Ag(100)–Cs, was there available definitive structural information on the metal–alkali interface which could contribute to the understanding of their selectivity and activity. We shall see that subsequently LEED evidence was obtained for surface restructuring in the presence of alkali metals, with more recently STM providing atom-resolved evidence.

Since much of the impetus for our STM studies stems from earlier spectroscopic investigations of alkali metals and alkali metal-modified surfaces,[6] we consider first what was learnt from the caesiated Cu(110) surface concerning the role of different oxygen states, transient and final states, in the oxidation of carbon monoxide, and then examine how structural information from STM can relate to the chemical reactivity of the modified Cu(110) surface.

6.2 Infrared Studies of CO at Cu(110)–Cs

At 80 K, carbon monoxide adsorbs at Cu(110) with a loss-peak in the vibrational spectrum at 2085 cm^{-1}; when this is exposed to Cs at 80 K, the loss-peak moves down in frequency to 1730 cm^{-1} and on warming to 298 K two vibrational peaks are present[6] at 1450 and 1960 cm^{-1}. In the absence of caesium, Cu(110) does not adsorb carbon monoxide at 298 K. Vibrational loss features associated with adsorbed carbon monoxide are also observed at 1430, 1600 and 1800 cm^{-1} at higher temperatures (360 K) for a Cu(110)–Cs surface where the Cs concentration is 3.6×10^{14} cm^{-2}. Very similar electron energy loss spectra were reported for potassium covered Ru(0001) and Pt (111) surfaces and discussed in detail by Bonzel and Pirug.[2] A coadsorbed K$^{\delta-}$–CO$^{\delta-}$ surface complex with direct or indirect electronic interaction (including electron transfer between K and CO) was considered by Solymosi and Berkó[7] to be the most likely explanation for the vibrational loss features. Complexes such as K$_x$–CO$_y$ were rejected.

However, in CO oxidation studies at Cu(110)–Cs, the formation of carbonate was observed[6] with features characteristic of bidentate (1500 cm^{-1}) and monodentate (1380 cm^{-1}) structures (Figure 6.2). It is the significance and specificity of the oxygen state present at the Cu(110)–Cs surface that is central to the oxidation chemistry observed. There are clearly analogies to be made with Iwasawa's study of CO oxidation at low temperatures, ammonia oxidation and water oxidation (see Chapter 5), with the highest reactivity associated with transient oxygen states which have not attained the thermodynamically stable O^{2-} state. What, then, can we learn from structural studies (LEED and STM) regarding oxygen states at a Cu(110)–Cs surface?

6.3 Structural Studies of the Alkali Metal-modified Cu(110) Surface

6.3.1 Low-energy Electron Diffraction

Tochihara and Mizuno,[8] Over,[9] Diehl and McGrath[10] and Barnes[11] have reviewed structural aspects of alkali metals present at metal surfaces. Over[9] collated surface crystallographic data from a wide range of experimental methods, LEED, SEXAFS, NEXAFS, SXRD and ion scattering; STM was considered to be of minor importance in the context of the article and to be more significant "in the study of defects and kinetics" at alkali metal-modified surfaces. Barnes[11] reviewed the early LEED studies of fcc metals modified by alkali metals with the evidence for missing row (1 × 2) and (1 × 3) structures seen to be a very general phenomenon. The alkali metal atoms are located within the missing rows with maximum separation of the atoms along the troughs due to the strong dipole–dipole repulsion. Marbrow and Lambert[12] and Hayden *et al.*[13] studied Cs adsorption at Ag(110), with the latter suggesting induced surface reconstruction with similar arguments to those suggested for Cu(110).

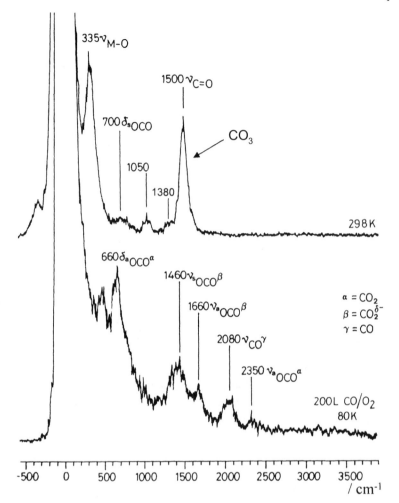

Figure 6.2 VEEL spectra when a mixture of CO and O_2 was coadsorbed at a Cu(110)–Cs surface ($\sigma_{Cs} = 3.5 \times 10^{14}$ cm^{-2}) at 80 K and the adlayer warmed to 298 K. Note the formation of surface carbonate (cf. Figure 6.9). (Reproduced from Ref. 6).

A series of LEED intensity studies, together with ion-scattering spectroscopy, established that a missing row structure was the correct model for the (1 × 2) phase,[14] with some small subsurface relaxation and reconstruction.[10]

6.3.2 Scanning Tunnelling Microscopy

Atom resolved studies were first reported by Ertl's group[15] in the early 1990s for Cu(110)–K, indicating the development of (1 × 3) and (1 × 2) structures depending on the surface coverage. They were missing row structures with every third row of the copper substrate atoms missing in the (1 × 3) structure, with the alkali metal atoms occupying the troughs. Even at the lowest coverage investigated, single K atoms were shown to squeeze copper atoms out of the [1$\bar{1}$0] rows, the process being driven by the high coordination of the K atoms within the holes and therefore a greater chemisorption energy which more than compensates for the formation of the hole and the di- or tri-vacancy left at the copper (1 × 1) surface. The authors concluded that the step-edges running along the <100> direction are formed by Cu(110) rows and not by the potassium atoms which LEED had established were mobile at room temperature. The Cu(110) rows are imaged as protrusions and the potassium filled the troughs and imaged as indentations; the troughs are 0.5 Å deep. The invisibility of the potassium atoms in the STM images is attributed to them being embedded rather than located on top of the close-packed metal surface. Less attention was given to the Cu(110)–Cs system.

6.3.3 Cu(110)–Cs System

With the advantage of *in situ* XPS, the images observed as a function of the caesium surface concentration could be followed, the latter calculated from the areas of the Cu($2p_{3/2}$) and Cs($3d$) peaks in the photoelectron spectrum.[16] The binding energy of the Cs($3d_{5/2}$) peak was centred at 725.6 eV and was invariant with coverage. Images corresponding to increasing caesium concentrations at 295 K (1.5×10^{14}, 1.9×10^{14} and 2.1×10^{14} cm^{-2}) are shown Figure 6.3a–c. At the lowest caesium concentration, a coverage of 10% of the copper atom monolayer, the surface is poorly imaged but some islands of local order are observed and also badly ordered structures in the <1$\bar{1}$0> direction.

Increasing the caesium concentration to 1.3×10^{14} cm^{-2} leads to improved ordering and evidence for two distinct structures (Figure 6.4). The structure in the lower right-hand side of the image consists of rows in the <1$\bar{1}$0> direction separated by 1.1 nm (three times the unreconstructed copper lattice spacing). This is the "low-coverage" (1 × 3) structure first reported by LEED. The other structure present at this caesium concentration has not been reported in previous studies of the system but is similar to that observed at lower caesium concentrations, *e.g.* 0.85×10^{14} cm^{-2}. The accompanying line profile (Figure 6.3) indicates maxima in the <100> direction of 0.72 nm, which is in registry with the substrate copper atoms, and 0.36 nm in the <110> direction, which is not. There are irregular height differences in the <1$\bar{1}$0> direction with a maximum amplitude of 0.02 nm. The structure can be described by a pseudo square lattice of side length 0.5 nm rotated by 45° to the substrate. The concentration of the atoms giving rise to the observed intensity maxima in the line profile is far in excess of the Cs concentration calculated from the Cs($3d_{5/2}$) intensity, which suggests that it is the copper atoms that are responsible for the maxima, although we cannot rule out an abnormally high local concentration of Cs. The pseudo square lattice does not occur at higher Cs concentrations and at 1.5×10^{14} Cs atoms cm^{-2} the surface is dominated by the

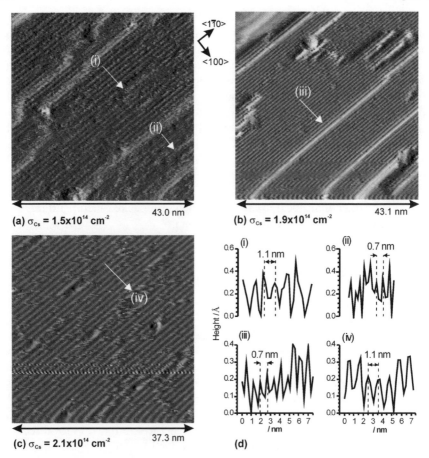

Figure 6.3 (a) Coexisting (1×3) and (1×2) structures at Cu(110), $\sigma_{Cs} = 1.5 \times 10^{14}\,\text{cm}^{-2}$; (b) (1×2) structure at $\sigma_{Cs} = 1.9 \times 10^{14}\,\text{cm}^{-2}$; (c) "high-coverage" (1×3) structure at $\sigma_{Cs} = 2.1 \times 10^{14}\,\text{cm}^{-2}$. Line profiles for each structure also shown. (Reproduced from Ref. 16).

(1×3) row structure with the 1.1 nm inter-row spacing (Figure 6.3a). This structure does, however, coexist with regions where the rows are separated by only 0.7 nm (twice the substrate copper lattice spacing), which is the only structure present when the concentration of caesium is increased to $1.9 \times 10^{14}\,\text{cm}^{-2}$ (Figure 6.3b). The (1×2) structure transforms with further caesium adsorption to give the "high-coverage" (1×3) structure (Figure 6.3c).

6.3.4 Oxygen Chemisorption at Cu(110)–Cs

Oxygen chemisorption at caesiated Cu(110) indicates facile surface mobility and structural transformations[16] at 295 K. For a caesium concentration of

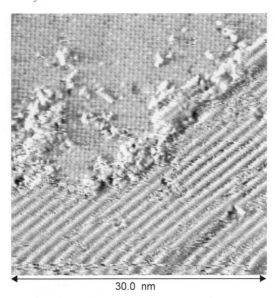

Figure 6.4 Coexisting pseudo square and "low caesium coverage" (1 × 3) structures at Cu(110); $\sigma_{Cs} = 1.3 \times 10^{14}$ cm^{-2}. (Reproduced from Ref. 16).

1.5×10^{14} cm^{-2} exhibiting the "low-coverage" (1 × 3) structure, both (2 × 1) and (3 × 1) oxygen-induced structures running along the <100> direction (*i.e.* at right-angles to the caesiated structures) develop after an exposure of 42 L (Figure 6.5b). After 20 L exposure only the (2 × 1) structure is present (Figure 6.5a). The oxygen concentration calculated from the O(1s) intensity is 3.2×10^{14} cm^{-2}. In addition to these two structures, there is evidence for parts of the surface developing a c(6 × 2) structure (Figure 6.5b and d), which is usually associated with defective oxygen states and analogous to what has been reported for Cu(110) either when it is exposed to high oxygen pressures[17] or the thermal activation of a disordered oxygen adlayer[18] at 80 K after warming to 295 K.

A real-time study of oxygen chemisorption at a caesiated copper surface ($\sigma_{Cs} = 1.5 \times 10^{14}$ cm^{-2}) illustrates how the original (1 × 3)Cs structures with rows running in the <110> direction gradually transforms to its "final state" structure, with new terraces emerging and (2 × 1)O structures running in the <100> direction (Figure 6.6), the black dots locating identical points on the surface in each image.

These images have significant implications for developing models of catalytic oxidation reactions in that *the surface is undergoing facile structural changes with transient sites being generated.* With time, these decay to give the final state with structures (rows) running at right-angles to those present in the original Cu(110)–Cs surface. The nature of the reactive oxygen state present *during the structural transformation* and active in the formation of $CO_2^{\delta-}$ and carbonate during CO oxidation at a Cu(110)–Cs surface has, however, not been isolated.

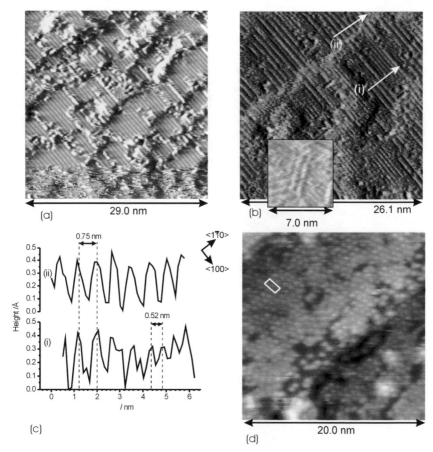

Figure 6.5 Oxygen chemisorption at a Cu(110)–Cs surface at 290 K. Image (a) is after 20 L oxygen exposure with a (2 × 1) structure present; image (b) is after 42 L oxygen exposure with both (2 × 1) and (3 × 1) states present; line profiles of the rows running in the <100> direction also shown, inter-row spacings are twice and three times the Cu–Cu distance in the <110> direction (c). Also shown is the image of a c(6 × 2) structure present as a minor component (b, d). (Reproduced from Ref. 16).

Adsorption of caesium at a Cu(110)–O surface exhibits different structural features (Figure 6.7). First there are well-ordered c(2 × 4) domains, but second, and somewhat unusual, they coexist with features which have no simple relationship with either the Cu(110) substrate atoms or the c(2 × 4) structure.[16] The rows are "bent", suggestive of strained structures associated with the (2 × 1)O rows modified by caesium adsorption.

The interatomic spacing within the rows of the c(2 × 4) structure is 0.5 nm, which is close to the Cs–Cs spacing in the monolayer of Cs formed at a Cu(110) surface at 80 K. The presence of the oxygen adlayer apparently prevents reconstruction of the surface with the caesium "locked in" within the rows of

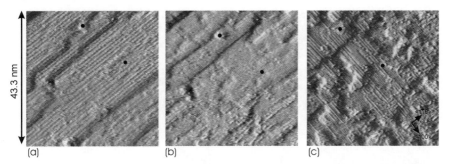

Figure 6.6 Real-time images of oxygen chemisorption at 290 K at a Cu(110)–Cs overlayer, $\sigma_{Cs} = 1.5 \times 10^{14}\,\mathrm{cm}^{-2}$. The black dots identify identical surface positions in the three images. Note the transformation of the surface from rows running in the <110> direction to rows running in the <100> direction and also changes in the step structure. (Reproduced from Ref. 16).

the Cu(110)–O adlayer. This is clearly seen[16] in image (f) after annealing the Cu(110)–O + Cs adlayer present at 295–500 K.

The composite surface adlayer of caesium coexisting with the ordered (2 × 1)O structures becomes clear by varying the tunnelling conditions. In images (c) and (e), the caesium overlayer structure is observed, whereas in image (d) the (2 × 1)O overlayer stands out. Line profiles in images (a) and (f) are shown in (g); in the <100> direction rows are separated by approximately 0.5 nm, resulting in every other row being out of registry with the underlying (2 × 1)O structure. This is more obvious in the highlighted part of the image shown in image (f).

6.4 Reactivity of Cu(110)–Cs to NH_3 and CO_2

The oxygen states present at the Cu(110)–Cs surface ($\sigma_0 = 1.6 \times 10^{14}\,\mathrm{cm}^{-2}$) were unreactive to ammonia at 295 K both for low caesium coverage ($\sigma = 1.5 \times 10^{14}\,\mathrm{cm}^{-2}$) and high coverage ($\sigma = 3.2 \times 10^{14}\,\mathrm{cm}^{-2}$), no changes being observed in either the XP spectrum or STM images.[16] However, exposure to ammonia at 475 K resulted in an N(1s) peak at 397 eV, consistent with the formation of N(a) and NH(a) states of total concentration $2 \times 10^{14}\,\mathrm{cm}^{-2}$. Although the surface is largely disordered, there is some evidence in the STM image of rows running in the <110> direction characteristic of NH_x species observed in previous studies of the Cu(110)–oxygen–ammonia system (see Chapter 5). The O(1s) spectrum at this stage indicates the presence of $1 \times 10^{14}\,\mathrm{cm}^{-2}$ of unreacted oxygen adatoms. This contrasts with the complete removal of oxygen at a Cu(110) surface at 475 K, reflecting the stronger binding of some of the oxygen states present at the Cu(110)–Cs surface.

Although XPS, HREELS and STM indicate that carbon dioxide is unreactive to the Cu(110) surface at 295 K, a caesium-modified surface ($\sigma = 1.5 \times 10^{14}\,\mathrm{cm}^{-2}$) results in the formation of well-ordered chain structures

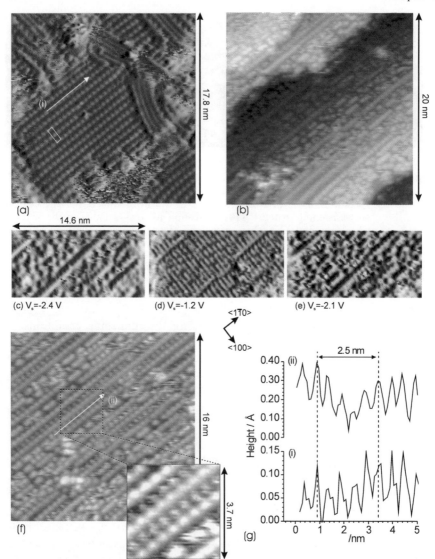

Figure 6.7 (a) Image of a low-coverage (2×1)O state at Cu(110) after exposure to caesium at 295K; $\sigma_0 = 1.7 \times 10^{14}\,\mathrm{cm}^{-2}$; $\sigma_{Cs} = 1.8 \times 10^{14}\,\mathrm{cm}^{-2}$; $c(2 \times 4)$ and "bent" chains form. (b) Adsorption of Cs on a monolayer of oxygen at 295 K. (c–e) Images of the same area as in (b) at different tunnelling conditions with (c) showing the Cs adlayer; (d) the (2×1)O adlayer; (e) recorded immediately after (d) showing the Cs adlayer; (f) after annealing (b) at 500 K (also inset); (g) line profiles from images (a) and (f). (Reproduced from Ref. 16).

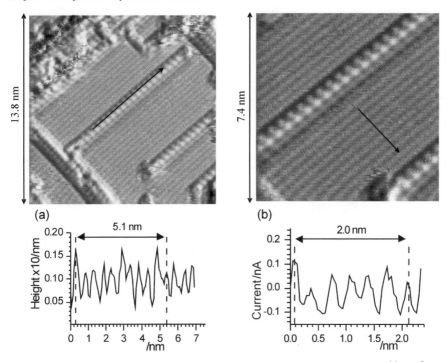

Figure 6.8 (a) Image of CO_2 chemisorption at Cu(110)–Cs ($\sigma_{Cs} = 1.4 \times 10^{14}$ cm^{-2}) at 295 K; well-ordered chains running in the <110> direction separated by the atom resolved structure of the Cu(110) surface with a spacing between the rows of 0.36 nm (see line profile). (b) The spacing within the chains is 0.51 nm (see line profile). i.e. close to twice the Cu–Cu distance within the copper rows running in the <110> direction. (Reproduced from Refs. 16, 18).

running in the <110> direction separated from each other by atom resolved unreconstructed copper atoms with a spacing of 0.36 nm (Figure 6.8). A line profile along the chains indicates a regular periodicity of 0.5 nm, which on the basis of the C(1s) binding energy at 288 eV and the energy loss spectra (Figure 6.9) is assigned to caesium carbonate. The loss-peak at 1510 cm^{-1} is characteristic of the bidentate structure, which is compatible with the STM chain structure observed with a periodicity of twice the Cu–Cu distance in the <110> direction. These observations are compatible with a chemisorption-induced surface reorganization[18] driven by the free energy of caesium carbonate formation and analogous to the self-reorganisation observed in the H_2–O_2 reaction at Rh(110)–K by scanning photoelectron microscopy.[19]

6.5 Au(110)–K System

Initial adsorption of potassium (0.15–0.25 ML) at room temperature does not change the (1 × 2) missing row structure of the Au(110) surface. The adsorbed

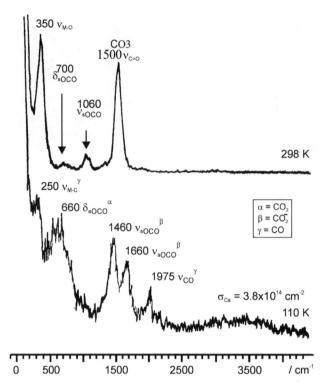

Figure 6.9 HREEL evidence for carbonate formation when CO_2 is chemisorbed at Cu(110)–Cs at 110 K and warmed to 298 K with a strong loss peak at 1500 cm^{-1} characteristic of a bidentate structure. (Reproduced from Ref. 6).

potassium atoms do not show up; furthermore, the K–K distance at a coverage of 0.15 ML corresponds to about 9 Å along [1$\bar{1}$0] in the reconstruction furrows If the K atoms did contribute, they should be observed in the STM image; the corrugation amplitude observed for this (1 × 2) structure is also typical of what is characteristic of the clean surface. At higher potassium coverages (e.g. 0.32 ML), the missing row (1 × 2) structure coexists with islands of c(2 × 2) (Figure 6.10). A model of this c(2 × 2) structure where potassium and gold atoms are localised to form an overlayer surrounded by (1 × 2) missing row areas is shown (Figure 6.10). Doyen et al.[20] indicate that the formation of the c(2 × 2) structure requires the gold atoms of the [1$\bar{1}$0] rows to be displaced by 4.08 Å (the Au lattice constant) in the [001] direction. The outermost atoms of the interrupted [1$\bar{1}$0] strings together with the displaced Au atoms form the c(2 × 2) unit cell. The vacancies in the [1$\bar{1}$0] strings are occupied by potassium atoms. The latter are situated 1.1 Å above the gold atoms and are imaged in the c(2 × 2) nucleus as depressions, whereas the gold atoms appear as protrusions. This, including the Cu(110)–K system, is characterised by strong theoretical support from a theory of STM that can account for the experimental data and

Surface Modification by Alkali Metals 115

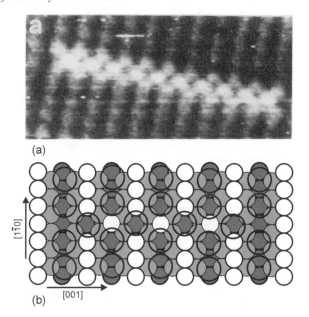

Figure 6.10 (a) Image of the c(2 × 2) nucleus formed at the Au(110)–K overlayer, $\theta_k = 0.33$ML. (b) Proposed model. (Reproduced from Ref. 20).

the electronic structure of the surfaces. Corrugation maxima and image inversion are shown to be the result of interaction between the STM tip and the surface.

6.6 Cu(100)–Li System

LEED data had already been analysed for the adsorption of lithium at Cu(100) at room temperature and indicated the development of (2 × 1), (3 × 3) and (4 × 4) structures with increasing coverage. The (2 × 3) LEED structure was interpreted as involving missing rows in the top layer of Cu(100) with lithium atoms present inside the troughs. The (3 × 3) and (4 × 4) structures were suggested to consist of Li adatoms and "substituted" Li atoms as shown in Figure 6.11. A rather special geometric structure was suggested by Mizuno et al.[21] involving lithium adatoms being trapped on small copper islands separated by "substituted" lithium atoms. In the STM investigation, the experimental set-up also included LEED facilities; the structures (2 × 1), (3 × 3) and (4 × 4) were therefore first confirmed by LEED and only then examined by STM. The (2 × 1) image (Figure 6.11) is noisy but there are obvious stripes running in the [$\bar{1}$10] direction; the distance between the stripes (5.1 Å) is about twice the copper–copper distance in the unreconstructed Cu(100) surface. The stripes reflect the copper rows and the lithium atoms are transparent in the image, an interpretation identical with that proposed for the Cu(100)–K

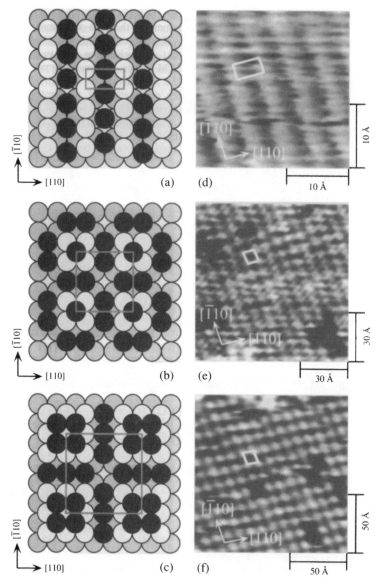

Figure 6.11 (a–c) Structures based on LEED studies of the adsorption of lithium at a Cu(100) surface with increasing coverage at 300 K. The (2×1), (3×1) and (4×4) structures are indicated; the black circles are lithium atoms, the grey circles are copper atoms in the sub-surface layer and the white circles are copper atoms in the mixed layer. (d–f) STM images of the corresponding (2×1), (3×3) and (4×4) structures. (Reproduced from Ref. 21).

system. Both the (3 × 3) and (4 × 4) images are less noisy but with distinct square arrangements of sides 7.5 and 10 Å, respectively, *i.e.* three and four times the Cu–Cu distance in the Cu(100) surface. The observed STM images of the lithiated surface at room temperature are therefore compatible with the "averaged" structures deduced from the earlier LEED analysis. The (2 × 1) is a missing row structure, while the (3 × 3) and (4 × 4) structures consist of both substitutional lithium atoms and lithium adatoms. The latter are present on small copper islands and imaged protrusions; they represent the first alkali metal atoms to be observed by STM to be present at metal surfaces. The (4 × 4) structure represents a very special arrangement of four lithium adatoms on islands of nine copper atoms separated by substitutional lithium atoms. The four lithium atoms are observed as a single protrusion and the protrusions form the (4 × 4) lattice, which is a relatively stable structure present on terraces. Accompanying its formation there was evidence in the STM images for the migration of vacancies and domains, which is the advantage STM has over LEED.

6.7 Summary

Some of the earliest LEED studies of the adsorption of alkali metals at metal surfaces were reported in the late 1960s by Gerlach and Rhodin.[22] At Ni(110), caesium adsorption led initially to streaking of the diffraction pattern in the [001] direction followed by sharp (1 × 3) and (1 × 2) patterns with the latter being dominant. Marbrow and Lambert[12] observed similar diffraction patterns for caesium adsorption on Ag(110), but it was Hayden *et al.*[13] at the Fritz Haber Institute in 1983 who, on the basis of LEED coupled with work function studies, suggested that the (1 × 2) phase was due to an alkali metal-induced surface reconstruction. Further investigations of alkali metal adsorption at Cu(110) established the generality of surface reconstruction and in 1989 Barnes *et al.*[14] with a quantitative $I(V)$ study of the Ag(110)–Cs and Cu(110)–K systems confirmed the missing row structures of the (1 × 2) phase.

What STM established first in 1991 for both Cu(110)–K and Cu(110)–Cs systems was the localised nature of the reconstruction process and the atom resolved details of the complexity of the structural changes observed with increasing coverage.[15] In 1993, Doyen *et al.*[20] provided theoretical support for the experimental observations with both the Cu(110) and Au(110) surfaces.

The influence of alkali metals on the vibrational frequencies of adsorbed carbon monoxide at a range of metals[2] was first reported in the early 1980s. Furthermore, oxygen states present at alkali metal-modified Cu(110) and Ag(110) were shown by HREELS and XPS to be unusually active in the oxidation of carbon monoxide[6] and hydrocarbons,[5] whereas the preadsorbed chemisorbed oxygen overlayer at these metal surfaces was uncreative under the same experimental conditions (see Chapters 2 and 5). What, then, could be learnt from STM regarding the active oxygen state present at alkali metal-modified Cu(110) surfaces?

At the Cu(110)–Cs surface, the structural changes observed with increasing caesium coverage indicated the development, in the following sequence, of $(1 \times 3) \rightarrow (1 \times 2) \rightleftharpoons (1 \times 3)$ phases with rows running in the <110> direction. When this surface is exposed to oxygen at 300 K, these are replaced by (2×1) and (3×1) structures with rows running in the <100> direction. The structural turbulence accompanying oxygen chemisorption (Figure 6.12) has obvious implications for catalytic oxidation reactions, where (for example) oxygen–carbon monoxide mixtures are exposed to the Cu(110)–Cs surface. The surface transformations leading to the final state will generate oxygen transients, the active sites, whose electronic structures have yet to be defined, but are characterised by exceptional oxidation activity.

Coupling STM with spectroscopic evidence (XPS and HREELS) has been shown to have obvious advantages in unravelling the oxidation chemistry of alkali metal-modified metal surfaces. What is also evident is that the surface structures observed are dependent on the sequence – alkali metal-modified surface exposed to oxygen or oxygen-modified metal surface exposed to the alkali metal! That doping oxides with alkali metals (lithium, potassium and sodium) resulted in partial oxidation of hydrocarbons, where the active oxygen state was O^-, was established by Lunsford and Lambert in the 1980s. References to these and related studies are given under Further Reading. At MgO, hydrogen abstraction from methane to generate CH_3 radicals occurs, with the activity associated with Li^+O^- centres. What is of interest is that coadsorption

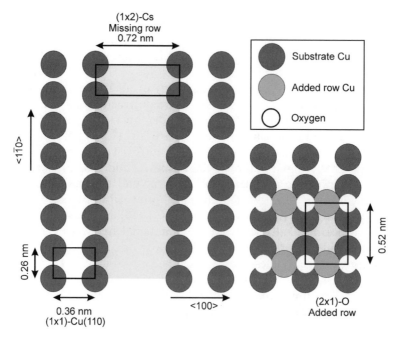

Figure 6.12 Structural transformations observed by STM when a Cu(110) surface is exposed to caesium followed by oxygen at 295 K.

studies (e.g. of ammonia and hydrocarbons) showed that at *atomically clean metal surfaces* (see Chapters 2 and 5) transient O^--like states were effective in H abstraction from a range of adsorbates at low temperatures; the activity, however, shuts down as O^{2-} states are formed as a result of surface reconstruction.

References

1. M. Bowker, in *The Chemical Physics of Solid Surfaces*, ed. D. A. King and D. P. Woodruff, Elsevier, Amsterdam, Vol. 6, 1993, p. 225.
2. H. P. Bonzel and G. Pirug, in *The Chemical Physics of Solid Surfaces*, ed. D. A. King and D. P. Woodruff, Elsevier, Amsterdam, Vol. 6, 1993, p. 51.
3. G. Ertl, M. Weiss and S. B. Lee, *Chem. Phys. Lett.*, 1979, **60**, 391.
4. A. Ozaki, K. Aiki and Y. Morikawa, in *Proceedings of the 5th International Congress on Catalysis*, ed. J. Hightower, North-Holland, Amsterdam, 1973, p. 1251.
5. A. F. Carley, A. Chambers, M. W. Roberts and A. Santra, *Isr. J. Chem.*, 1998, **38**, 393.
6. A. F. Carley, M. W. Roberts and A. J. Strutt, *Catal. Lett.*, 1994, **29**, 169; *J. Phys. Chem.*, 1994, **98**, 9175.
7. F. Solymosi and A. Berkó, *Surf. Sci.*, 1988, **201**, 361.
8. H. Tochihara and S. Mizuno, *Prog. Surf. Sci.*, 1998, **58**, 1.
9. H. Over, *Prog. Surf. Sci.*, 1998, **58**, 249.
10. R. D. Diehl and R. McGrath, *Surf. Sci. Rep.*, 1996, **23**, 43.
11. C. J. Barnes, in *The Chemical Physics of Solid Surfaces*, ed. D. A. King and D. P. Woodruff, Elsevier, Amsterdam, 1994, Vol. 7, p. 501.
12. B. A. Marbrow and R. M. Lambert, *Surf. Sci.*, 1976, **61**, 329.
13. B. E. Hayden, K. C. Prince, P. G. Davie, G. Paolucci and A. B. Bradshaw, *Solid State Commun.*, 1983, **48**, 325.
14. C. J. Barnes, M. Lindroos, D. J. Holmes and D. A. King, *Surf. Sci.*, 1989, **291**, 143.
15. R. Schuster, J. V. Barth, G. Ertl and R. J. Behm, *Phys. Rev. B*, 1991, **44**, 13689.
16. A. F. Carley, P. R. Davies, K. R. Harikumar, R. V. Jones and M. W. Roberts, *J. Phys. Chem.*, 2004, **108**, 14518.
17. D. Coulman, J. Wintterlin, J. V. Barth, G. Ertl and R. J. Behm, *Surf. Sci.*, 1990, **240**, 151; see also F. Besenbacher and I. Stensgaard, in *The Chemical Physics of Solid Surfaces*, Vol. 7, ed. D. A. King and D. P. Woodruff, Elsevier, Amsterdam, 1994, p. 573.
18. A. F. Carley, P. R. Davies and M. W. Roberts, *Philos. Trans. R. Soc. London, Ser. A*, 2005, **363**, 829.
19. H. Marbach, S. Gunther, L. Gregoratti, M. Kiskinova and R. Imbihl, *Catal. Lett.*, 2002, **83**, 161.
20. G. Doyden, D. Drakova, J. V. Barth, R. Schuster, T. Gritsch, R. J. Behm and G. Ertl, *Phys. Rev. B*, 1993, **48**, 1738.

21. S. Mizuno, H. Tochihara, Y. Matsumoto and K. Tanaka, *Surf. Sci.*, 1997, **393**, L69.
22. R. L. Gerlach and T. N. Rhodin, *Surf. Sci.*, 1968, **10**, 446; 1969, **17**, 32.

Further Reading

T. Ito and J. H. Lunsford, Synthesis of ethylene and ethane by partial oxidation of methane over lithium-doped magnesium oxide, *Nature*, 1985, **314**, 721.

D. J. Driscoll, W. Martir, J.-X. Wang and J. H. Lunsford, Formation of gas-phase methyl radicals over MgO, *J. Am. Chem. Soc.*, 1985, **107**, 58.

G. D. Moggridge, J. P. S. Badyal and R. M. Lambert, X-ray photoelectron spetroscopic characterisation of oxygen surface species on a doubly promoted manganese oxide model planar catalyst: significance for CH_4 coupling, *J. Phys. Chem.*, 1990, **94**, 508.

A. F. Carley, S. D. Jackson, J. N. O'Shea and M. W. Roberts, Oxidation states at alkali-metal doped Ni(110)–O surfaces, *Phys. Chem. Phys.*, 2001, **3**, 274.

M. Bender, O. Seiferth, A. F. Carley, A. Chambers, H.-J. Freund and M. W. Roberts, Electron, photon and thermally induced chemistry in alkali–NO coadsorbates on oxide surfaces, *Surf. Sci.*, 2002, **513**, 221.

CHAPTER 7
STM at High Pressure

"Having precise ideas often leads to a man doing nothing"

Paul Valéry

7.1 Introduction

One of the criticisms of experimental methods in surface science is that data obtained under ultra-high vacuum conditions could have little relevance to "real catalytic conditions" – the so-called "pressure gap" in catalysis. However, a worrying aspect of high-pressure studies is the inherent problem of ensuring that the gases are of very high purity (<1 ppm of impurities present). The problem is exacerbated if the reactants have low sticking probabilities and a potential contaminant is present of high sticking probability. At a pressure of 1 mbar, an impurity present at a concentration of 1 part in 10^6 and with a sticking probability of unity would form a surface monolayer in about 1 s at 295 K. A number of groups have described STM systems which can operate at high pressures; the first was Somorjai's group[1] at Berkeley in 1996, the second Besenbacher's group[2] in Aarhus in 2001 and more recently Freund's group[3] at the Fritz Haber Institute and Frenken's group[4] at Leiden.

The Aarhus group[2] laid down the ground rules for a successful STM experiment, particular attention being given to both substrate and gas purity. The authors describe in detail the design and performance of a high-pressure, high-resolution STM in a multipurpose UHV system (Figure 7.1). The main UHV chamber rests on an undamped steel frame; vibrational damping is not necessary due to the high eigenmode frequency spectrum of the STM. Auger electron spectroscopic (AES) and X-ray photoelectron spectroscopic (XPS) facilities are available together with a quadrupole mass spectrometer and a LEED system. The usual facilities for sample cleaning and sample transfer from the main chamber to the high-pressure cell are available. The STM is small and compact and mounted inside a heavy steel block, with all metal parts gold plated to avoid reactions with the gases being used. Fast scanning is possible with several constant current images being recorded per second with a

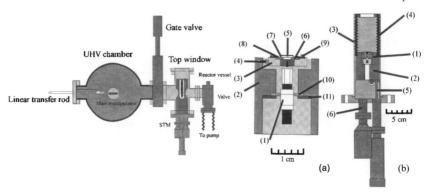

Figure 7.1 Reaction vessel containing the STM, the connection to the UHV system and the sample transfer mechanism: (a) and (b) are details of the high-pressure STM (a) and its mounting and assembly (b): (1) Inchworm scanner tip; (2) Invar housing; (3) sample holder; (4) quartz balls; (5) sample; (6) Ta support; (7) Ta foil; (8) leaf springs; (9) screws; (10) Macor ring; (11) support ring. (Reproduced from Ref. 2).

very low noise level. The STM experiments are limited to a pressure of about 1 bar. Great care is taken with the selection of the tip material and also in the preparation of the gases, where high purity is particularly critical for obtaining meaningful data at high pressures.

The authors chose two examples – hydrogen adsorption at Cu(110) and the hydrogen–Au(111) system – to illustrate the performance of the high-pressure STM system. The former is an example of a reactive gas whereas at Au(111) the hydrogen is unreactive.

It was established that STM images of Au(111) exhibited atomic images of the gold surface both in UHV and during an exposure to 0.85 bar of hydrogen at 300 K. Comparable corrugation amplitudes were observed in both cases using the same tunnelling conditions with also identical lattice constants. These experiments established that viewing the Au(111) surface is unaffected by the high pressure of a "reactive" gas (H_2); similar results were obtained with an inert gas, argon.

The authors then chose to examine hydrogen adsorption at Cu(110). This was a well-chosen example in that hydrogen adsorption is activated, being pressure dependent, and also was already known from LEED studies to exhibit a missing row (1 × 2) structure with every second close packed ($1\bar{1}0$) copper row missing at high hydrogen pressures. What, then, was learnt from STM?

A sequence of STM images were obtained (Figure 7.2) of the Cu(110) surface before hydrogen exposure (A), at an ambient H_2 pressure of 1 bar (B) and finally after evacuation under UHV conditions (C). It is clear that in the presence of H_2 the surface reconstructs into the well-known (1 × 2) missing row structure and that an evacuation the surface reconstruction is lifted with the (2 × 1) structure observed. AES established that no impurities were present at the Cu(110) surface.

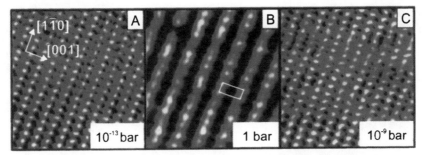

Figure 7.2 A sequence of STM images at 298 K for Cu(110) under UHV (A), 1 bar of H$_2$ (B) and after evacuation to 10^{-9} bar (C). Note the (1 × 2) missing row structure at 1 bar H$_2$ (B) and its reversibility on evacuation (C). (Reproduced from Ref. 2).

Some more detailed experiments at different hydrogen pressures established that the first evidence for the (1 × 2) missing row structure was obtained at a hydrogen pressure between 2 and 20 mbar (the precise value is not known). At 20 mbar, atom resolved images were not observed, only the (1 × 1) and (1 × 2) structures being seen. This is attributed to rapidly diffusing copper atoms, present during the formation of the (1 × 2) phase interfering with measurement and leading to the missing row structure. The high-pressure H$_2$–Cu(110) images also provided the first atom resolved imaging of a metal surface under a reactive gas at atmospheric pressure.

The authors also highlighted the significance that small quantities of impurities in the hydrogen can have. They exposed Cu(110) to hydrogen which was "pure" but had *not been further purified* by running it through a catalyst bed and molecular sieve. With increasing hydrogen exposure there developed p(2 × 1) and c(6 × 2) structures, which are typical of what is observed[2,5] with oxygen (but present in these experiments as an impurity). There was also evidence of a p(3 × 2) structure which is fully developed at 1 bar, which had not been reported previously; the periodicity was 0.38 nm in the ⟨110⟩ direction and 0.72 nm in the ⟨100⟩ direction (Figure 7.3).

7.2 Catalysis and Chemisorption at Metals at High Pressure

Somorjai's group has made outstanding contributions to the study of catalytic reactions at high pressures using STM. Recently, they have extended the capability of STM to be able to operate at up to 30 atm and to 600 K. Although much of what has been reported in the literature is based on an earlier system,[1] in 2006 a brief description was given of the new system with attention to overcoming the problems associated with detecting the products of a catalytic reaction in high pressures of background gases.[6] This was addressed by decreasing the volume of the reaction chamber to less than 10 cm^3 (compared

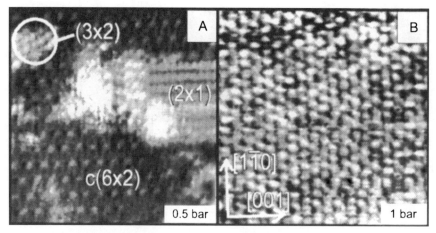

Figure 7.3 Impurity-induced structures on Cu(110): (A) three coexisting structures at an H_2 pressure of 0.5 bar; (B) fully developed (3 × 2) structure at an H_2 pressure of 1 bar; it is hexagonal with a periodicity of 3.8 Å in the [110] direction and 7.2 Å in the [001] direction. These images emphasise the need to purify the hydrogen rigorously. (Reproduced from Refs. 2,5).

with the original chamber of 10^4 cm^3) and replacing the batch reactor system with a flow cell so as to eliminate diffusion problems during reaction studies. The system was also made more robust to eliminate vibration problems during scanning. In view of the evidence that had been gleaned from surface science methods (LEED, XPS, HREELS, *etc.*) the Somorjai group was in a good position to compare their results and conclusions with what was revealed by high-pressure STM. They paid particular attention to how phase diagrams and molecular structure can be observed over wide pressure and temperature ranges. Some of the catalytic studies were reviewed[7] at a Discussion Meeting of the Royal Society in 2004 and the chemisorption studies[6] in 2006.

7.2.1 Carbon Monoxide and Nitric Oxide

The adsorption of carbon monoxide at Pt(111) was reported by both the Berkeley and Aarhus groups in 1998 and 2002, respectively.[8,9] In the pressure range 200–700 Torr at room temperature, Jensen *et al.*[8] reported the formation of a hexagonal structure with a periodicity of 12 Å, which Besenbacher and co-workers[9] established was the $(\sqrt{19} \times \sqrt{19})$ $R23.4°$ structure (Figure 7.4). Subsequently, Besenbacher's group in a detailed paper[10] in 2004 distinguished two different pressure ranges: below 10^{-2} Torr the "nearly hexagonal CO structure" exhibits a Moiré lattice vector oriented along a 30° high symmetry direction of the substrate corresponding to a pressure-dependent rotation of the CO overlayer with respect to the (1 × 1) Pt surface, whereas above 10^{-2} Torr the CO adlayer angle is independent of pressure. These observations are explained in terms of CO–CO repulsive interactions and the substrate potential.

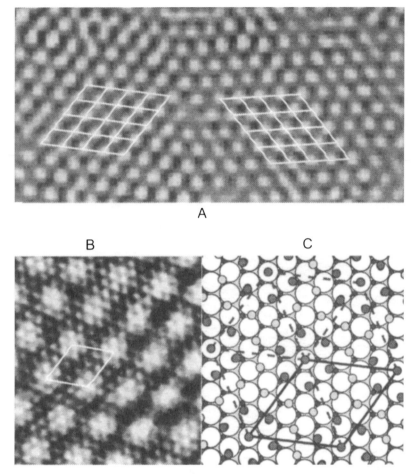

Figure 7.4 (A) STM image $(240 \times 125)^2$ of two rotational domains of the Moiré pattern formed at Pt(111) by CO at 1 bar at room temperature. (B) High-resolution image of the CO overlayer at 1 bar. (C) The $(\sqrt{19} \times \sqrt{19})$ R23.4°–13CO structure; the unit cell is shown; the dark circles represent CO molecules adsorbed in nearly atop sites. (Reproduced from Ref. 9).

At Rh(111) under the same conditions as for CO adsorption at Pt(111), the surface structures observed vary with the pressure. At the lowest pressure (10^{-8} Torr) a (2×1) overlayer forms but with increasing pressure the $(\sqrt{7} \times \sqrt{7})$ R19° appears in the 10^{-7}–1 Torr region, above which (up to 700 Torr) only a (2×2) structure is present[6].

Similar studies[6] were reported for the chemisorption of nitric oxide at Rh(111) with at 10^{-8} Torr a (2×2)–3NO structure, previously reported by LEED with a different structure with a (3×3) periodicity observed above 10^{-2} Torr (Figure 7.5). The latter has an apparent height of 0.1 nm above the (2×2) structure, suggesting that a distortion has occurred of the top surface layer of

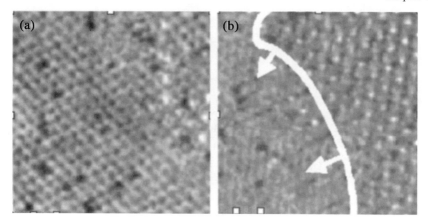

Figure 7.5 STM images of the chemisorption of nitric oxide at Rh(111) at a pressure of 0.03 Torr at 298 K showing the phase transition between (2×2) and a (3×3) structure. (a) $t = 0$ s; (b) $t = 110$ s; the phase boundary has now moved and is half-way across the image. (Reproduced from Ref. 6).

rhodium atoms. The transformation $(2 \times 2) \rightarrow (3 \times 3)$ is rapid at room temperature with an energy barrier of about 0.72 eV.

7.2.2 Hydrogenation of Olefins

There have been, other than by Somorjai's group, very few studies of *catalytic* reactions followed by STM at high pressures. The hydrogenation of olefins on metal surfaces is one of the most extensively studied catalytic reactions, but recent studies by Somorjai's group have established[7,11] how high-pressure studies with atomic resolution of the surface has emphasised the crucial role of surface mobility in maintaining catalytic activity. Key intermediates in ethylene hydrogenation are ethylidene (HC–CH$_3$) and ethylidyne (C–CH$_3$). At Pt(111), ethylidyne remains stable up to 430 K when further loss of hydrogen and C–C bond cleavage occur with the formation of C$_2$H (acetylide). These species were confirmed from their vibrational spectra.

When hydrogen is introduced in excess in the 1–100 Torr range along with ethylene at 295 K, the dehydrogenation reaction slows such that only σ-bonded ethylene and ethylidyne coexist at the Pt(111) surface. Under these conditions, ethylene hydrogenation occurs with a turnover rate of about 10 ethane molecules produced per platinum atom per second. Under high-pressure conditions of hydrogen and ethylene, ethylidyne remains strongly adsorbed on the platinum surface for about 10^6 turnovers (molecules of ethane formed per platinum surface per second). Hydrogenation occurs through weakly adsorbed π-bonded species that occupy atop sites; these species hydrogenate sequentially to ethyl and then to ethane. Such sites are available on all crystal planes of platinum and thus explain the lack of structure sensitivity.

STM at High Pressure

STM studies indicate that ethylidyne is mobile on the Pt(111) surface at 300 K at both low and high ethylene pressures with or without hydrogen. Somorjai and Marsh[7] emphasise that for the reaction to occur at high pressures, when the surface is close to saturation coverage, mobility within the adlayer must be maintained in order to make available active sites for the catalytic reaction. This was tested by adding carbon monoxide; for both Pt(111) and Rh(111) surfaces catalytic activity was immediately halted when CO was coadsorbed. STM indicates that under these conditions the previously mobile surface species (Figure 7.6a and b) became locked-in into static ordered structures (Figure 7.6c and d). Surface mobility is suppressed with the availability of reactants to reach active sites inhibited. The restriction of surface

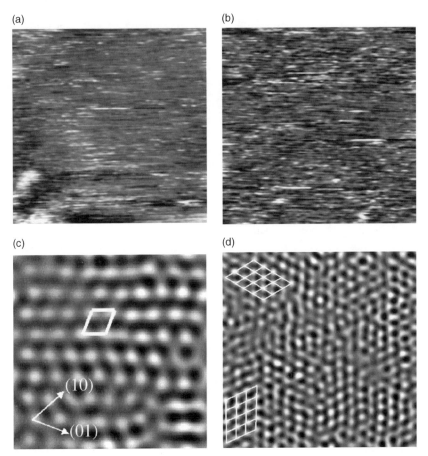

Figure 7.6 STM images $(100 \times 100)\,\text{Å}^2$ of Pt(111) under different catalytic conditions:[7] (a) 20 mTorr H_2; (b) 20 mTorr H_2 and 20 mTorr C_2H_4; (c) 20 mTorr H_2 plus 20 mTorr C_2H_4 and 2.5 mTorr CO(g). The CO added induced the formation of a $(\sqrt{19} \times \sqrt{19})$ R23.4° structure in (c). In (d) are shown two rotational domains of the $\sqrt{19}$ structure. (Reproduced from Ref. 7).

mobility appears to be a frequent occurrence in coadsorption, a further example being sulfur adsorbed at Cu(110) at 295 K on exposure to oxygen.[12] In this case, the mobile sulfur adatoms became locked in to a c(2 × 2) structure separated from each other by short (2 × 1)O chains. These oxygen states are expected to be more active in oxydehydrogenation reactions than the well-ordered layer (2 × 1) domains associated with the Cu(110)–O system. Under such circumstances, sulfur could be viewed as a "promoter".

Analogous studies to those of ethylene were also reported by the Somorjai group for cyclohexene at Pt(111).[11] At pressures less than 1 Torr at 300 K the molecule undergoes partial dehydrogenation to form the π-allyl species C_6H_9, but which undergoes further dehydrogenation as the temperature is increased. The chemistry and surface structures observed by STM are very dependent on the hydrogen-to-cyclohexene ratio used. With a 10:1 mixture at room temperature the surface is disordered with high mobility of the adsorbate and both benzene and cyclohexane are products. Addition of carbon monoxide to the mixture brings the catalysis to a halt and the STM image indicates the characteristic $(\sqrt{19} \times \sqrt{19})$ $R23.4°$ structure of a chemisorbed carbon monoxide. With the 10:1 mixture at 350 K the surface is again disordered but the major product is benzene; again, CO deactivates the surface but STM indicates that no ordered structure is present when catalytic activity is observed (Figure 7.7).

In Leiden, Hendriksen et al.[4] have recently described a "reactor STM" where information on surface structure from STM is coupled with kinetic studies by on-line mass spectrometry. Studies of CO oxidation, under flow "reactor" conditions, at Pt(110) and Pd(100) surfaces over a range of temperature and high pressures (>1 bar) were reported. With oxygen-rich conditions the surface, with both metals, was very active in CO_2 formation, the "active oxygen" being suggested to be extracted from the "oxide" surface under the dynamic conditions used. Langmuir–Hinshelwood behaviour, observed when the preadsorbed oxygen state present at the metal surface is exposed to CO(g),

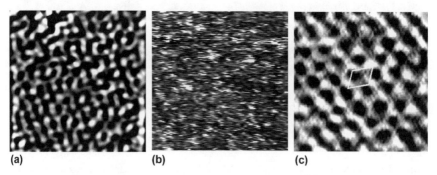

(a)   (b)   (c)

Figure 7.7 STM images of Pt(111) at 300 K: (a) (75 × 75) Å^2, 20 mTorr cyclohexene plus 20 mTorr H_2; no catalytic products formed; (b) (50 × 50) Å^2, 200 mTorr H_2, 20 mTorr of cyclohexene, disordered surface and cyclohexane formed; (c) (90 × 90) Å^2, CO added, no catalytic activity. (Reproduced from Ref. 11).

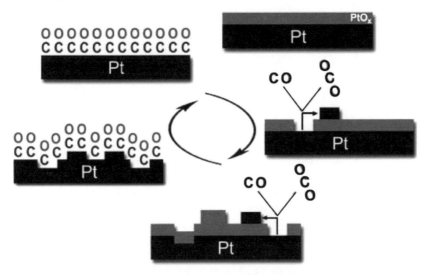

Figure 7.8 Schematic representation[4] of mechanism of CO oxidation at Pt(110). At low partial pressure of oxygen the surface is almost completely covered by CO(a); reaction is by CO(a) and O(a) by a Langmuir–Hinshelwood mechanism. At high oxygen pressure an oxide overlayer forms; reaction now takes place between CO(g) and "oxide" atoms removed from the oxide overlayer leading to a "rough" oxide surface. At sufficiently high CO pressure the oxide is completely removed, leaving a rough metallic Pt(110) surface, with strong analogies with the ammonia–oxygen reaction at Cu(110) (see Chapter 5). (Reproduced from Ref. 4).

does not hold; the behaviour is suggested to be more akin to the Mars–van Krevelen mechanism (Figure 7.8). This is reminiscent of what emerged from detailed catalytic oxidation studies of a variety of adsorbates (NH_3, CO, H_2O, CH_3OH, *etc.*) by the coadsorption approach and later by STM and symptomatic of "active" transient oxygen states present which are distinct from the oxide-type O^{2-} states. It is interesting that the authors also observe that the oxide adlayer becomes increasingly rough, supporting the view that it is the oxygens at the "edges" that are the reactive ones, as was suggested for ammonia oxidation at a Cu(110)–O surface (Chapters 2, 4 and 5).

7.3 Restructuring of the Pt(110)–(1 × 2) Surface by Carbon Monoxide

The chemisorption of CO at Pt(110) is one of the most extensively studied systems, which exhibits structural transformation induced by an adsorbate, with most experimental methods available in surface science being used. It was, however, the Aarhus group that provided atom resolved evidence, over the pressure range 10^{-9}–10^3 mbar and temperature range 300–400 K, for a simple model that explained the observed transformations.[13] The essential feature of the model is that the creation of low-coordinated platinum atoms with increasing CO coverage results in the binding energy of the CO increasing linearly with the decrease in the coordination number of the platinum. This conclusion is supported by density functional theory calculations which show that CO bonds most strongly to low-coordination metal atoms.

The overall model is illustrated by the step-density plot as a function of surface coverage; a step atom is defined as an atom with a coordination number of 5 or 6. Previous classical studies have been interpreted in terms of the difference between the adsorption energies of carbon monoxide at the (1×2) and (1×1) clean platinum surfaces and binds by about 0.45 eV more strongly to a 5-coordinated atom than to a 7-coordinated atom. This is about the energy required to break two Pt nn bonds and is the explanation for how CO creates its own low-coordinated adsorption site and hence the observation of the lifting of the (1×2) reconstruction. What is evident from the Aarhus study is that no new phase was observed for CO adsorption at Pt(110) by bridging the pressure gap, *i.e.* increasing the pressure to 1 bar. The authors also comment on the earlier pioneering high-pressure studies of the Berkeley group, where atomic resolution was not achieved but large flat terraces displaying no missing row reconstructions were observed. They proposed that surface cleanliness may account for the differences observed by the two groups, a problem inherent to high-pressure studies and the reason why, in our high oxygen pressure XPS studies[14] in 1979, we chose to study an sp-metal (silver) as a substrate, known not to be highly reactive to potential contaminants in UHV systems. The sensitivity of reaction pathways to the presence of oxygen as a contaminant was further illustrated[15] by studies of carbonate formation at a polycrystalline copper surface when various oxygen–carbon dioxide mixtures were exposed to it at room temperature. For mixtures containing 300 ppm of oxygen, the surface was "oxidised" with no evidence of carbonate; however, when oxygen was present at ≤ 70 ppm surface carbonate was the dominant surface species present. With "high-purity" CO_2 no reaction occurred at the Cu(110) surface at room temperature.

The following are the steps involved, with (4) being a highly efficient scavenging reaction leading to surface carbonate, with $CO_2(s)$ being in equilibrium with $CO_2(g)$ and present at immeasurably low concentration and undergoing surface 'hopping'.

$$CO_2(g) \rightarrow CO_2(s) \quad \text{carbon dioxide undergoing surface diffusion} \quad (1)$$

$$O_2(g) \rightarrow 2O^{\delta-}(s) \quad \text{dissociative chemisorption of dioxygen} \quad (2)$$

$$O^{\delta-}(s) \rightarrow O^{2-}(a) \quad \text{"oxide" formation : unreactive to } CO_2 \quad (3)$$

$$CO_2(s) + O^{\delta-}(s) \rightarrow CO_3(a) \quad \text{low-energy pathway to carbonate} \quad (4)$$

7.4 Adsorption-induced Step Formation

Step formation at an atomically clean metal surface involves the breaking of metal–metal bonds and is usually associated with high temperatures. However, if the metal has chemisorbed species present, the metal–metal bonds will be weakened and so metal atom mobility would be facilitated and occur at very much lower temperatures. In the Pt(110)–CO system, the energy required to break Pt–Pt surface bonds is suggested to approach zero when more than 50% of the atoms are associated with steps.[13,16] It is suggested that such adsorption-induced microroughening of the surface is a general phenomenon and may determine the surface reactivity of metal surfaces at high gas pressures. There is, however, STM evidence that such roughening can also occur at low temperatures and low pressures when the surface coverage is high and therefore simulates the high-pressure studies at high temperatures.

7.5 Gold Particles at FeO(111)

Although there have been extensive studies of gold model systems under UHV conditions, it was Freund and his colleagues at the Fritz Haber Institute in Berlin who addressed the question of whether morphological changes could be induced at high gas pressures.[3] There were two factors which could be of general concern: first a weakening of the metal–metal bonds within the particle and second a decrease in the energy of interaction between the gold particle and the support. Furthermore, the support itself may react with the reactive gas with consequences for particle–support interaction. The authors provide in the introduction to their paper a good review of gold particles on substrates; these are considered here in Chapter 9.

The high pressure STM studies focus on the Au–FeO(111) system.[3] The FeO support has certain advantages in that it is free of vacancies and line defects, which may have a strong influence on both structure and reactivity. In this way, the authors isolated any morphological changes that might be observed to the interaction between the gases and just the gold particles. The STM is housed in a chamber separated from the main chamber in order to obtain STM images in various gases. There is also a further gold-plated high-pressure cell attached to the main chamber, which enables gas exposures to be made similar to those in the STM chamber itself. This allows any changes in the morphology by STM through being induced by tip changes at high pressures to be ascertained. The FeO(111) surface had been established to be unaffected by gases. Great care was taken in gas purification and also to ensure that (when CO was used) no carbonyl formation occurred, which would "contaminate" the gold surface.

Detailed movie STM studies are reported with carbon monoxide, oxygen, carbon monoxide–oxygen mixtures and hydrogen in the pressure range 10^{-6}–2 mbar. In general, the gold particles were fairly stable in both oxygen and hydrogen up to 2 mbar; however, in the presence of carbon monoxide and carbon monoxide–oxygen mixtures changes in the morphology of particles

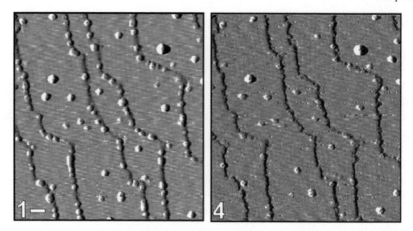

Figure 7.9 Two snapshots of Au particles on FeO(111) with increasing CO pressure: (1) 10^{-6} Torr; (4) 2 mbar CO; note the removal of gold particles from step-edges. (Reproduced from Ref. 3).

present at step-edges were observed (Figure 7.9). The authors also emphasised that observations attributed to CO dissociation at gold particles are likely to be incorrect and due to gas impurity-driven morphological changes.

7.6 Hydrogen–Deuterium Exchange and Surface Poisoning

The H_2–D_2 exchange reaction has been one of the most significant reactions used to study the catalytic activity of a surface and evidence for the dissociative chemisorption of hydrogen. It is, therefore, also a probe of the efficacy of how or why a surface's activity can be poisoned. Somorjai's group at Berkeley has reported[17] how carbon monoxide can control the exchange reaction at Pt(111) over wide temperature and pressure ranges – mTorr to atmospheric pressure. They have used high-pressure STM, SFG and XPS, each technique housed in separate systems, and therefore different from the Cardiff STM system, which had *in situ* XPS for chemical information, to complement directly the observed topographical data.

The Berkley study has far-reaching consequences for a detailed understanding of heterogeneously catalysed reactions in that the authors have delineated the precise conditions under which the presence of CO(g) can inhibit completely the exchange reaction, how increasing the temperature results in the reaction being initiated, with surface mobility and vacancy formation being central to maintaining a high rate of exchange.

At room temperature, no exchange is observed in the presence of 200 mTorr H_2, 20 mTorr D_2 and 5 mTorr CO. Under these conditions, a monolayer of ordered CO was observed by STM. It is a hexagonal ordered structure incommensurate with respect to the Pt(111) surface and a coverage of about

STM at High Pressure 133

0.6 monolayers. This structure is similar to that observed for pure CO at Pt(111). At 345 K, no periodic structure was observed by STM indicative of the mobility of the CO(a) and the exchange reaction occurred as if no CO was present.

In the absence of CO(g), the exchange reaction was fast at room temperature and STM indicated the adlayer to be disordered. We therefore have a further example of where surface disorder can be correlated with high catalytic activity. Other examples such as in oxidation catalysis are discussed elsewhere (Chapter 5).

The Berkeley results also emphasise the novel observations of Salmeron that for the dissociation of H_2 to occur readily at Pt(111) three or more vacancies are a prerequisite (Chapter 8). The vacancies also encourage surface mobility (disorder), now recognised as a significant factor in determining catalytic activity.

7.7 Summary

Although both LEED and surface spectroscopies (AES and XPS) provided evidence on both the structure and the atomic composition of surfaces, they have been limited in the main to ultra-high vacuum pressure conditions. STM studies at high pressure have been relatively few, but the atom resolved evidence indicates that subtle but significant changes in both surface structure and kinetic behaviour occur with increasing pressure of the reactants. A theme that is common to many of the systems investigated is that surface mobility of both adatoms and substrate (metal) atoms, including step-movement, is a common phenomenon. The observed correlation of chemical reactivity with surface mobility is discussed in Chapters 2, 4 and 5.

Hydrogen chemisorption resulting in the reconstruction of a Cu(110) surface, is shown to be pressure dependent and reversible resulting in the (1×2) missing row structure. Chemisorption structures of carbon monoxide at Pt(111) are pressure dependent, with distinction made between those observed below 10^{-2} Torr (a nearly hexagonal structure) and those at higher pressures. At Rh(111), carbon monoxide adsorption results in a (2×1) overlayer at 10^{-8} Torr, a $(\sqrt{7} \times \sqrt{7})$ $R19°$ up to 1 Torr and a (2×2) structure up to 700 Torr. Changes were also reported for nitric oxide chemisorption at Rh(111) as the pressure was increased from 10^{-8} to above 10^{-2} Torr, with the suggestion that there is also a distortion of the surface rhodium atoms.

The Pt(110)–carbon monoxide system has been thoroughly studied with a model developed where the creation of low coordinated platinum atoms controls the binding energy (the heat of chemisorption) of the carbon monoxide. Two groups (Berkeley and Leiden) have given attention to catalytic reactions, hydrogenation of ethylene, H_2–D_2 exchange and CO oxidation. Surface mobility is shown to be a prerequisite of catalytic activity; the addition of CO to the reactants, however, induces a "static, ordered structure", exhibiting no activity.

A frequent theme in high-pressure STM is the possibility of "artefact structures" developing due to low levels of impurities present in the gaseous reactants; it is an aspect that has been discussed[13] for the Pt (110)–CO system.

References

1. G. A. Somorjai, *Faraday Discuss.*, 1996, **105**, 263; in *Dynamics of Surfaces and Reaction Kinetics in Heterogeneous Catalysis*, ed. G. C. Froment and K. C. Waugh, Elsevier, Amsterdam, 1997, 35.
2. E. Laegsgaard, L. Osterlund, P. Thostrup, P. B. Rasmussen, I. Stensgaard and F. Besenbacher, *Rev. Sci. Instrum.*, 2001, **72**, 3537.
3. D. E. Starr, S. K. Shaikhatinov and H. -J. Freund, *Top. Catal.*, 2005, **36**, 33.
4. B. L. M. Hendriksen, S. C. Babaru and J. W. M. Frenken, *Top. Catal.*, 2005, **36**, 43.
5. L. Osterlund, P. B. Rasmussen, P. Thostrup, E. Laesgaard, I. Stensgaard and F. Besenbacher, *Phys. Rev. Lett.*, 2001, **86**, 460.
6. M. Montano, D. C. Tang and G. A. Somorjai, *Catal. Lett.*, 2006, **107**, 131.
7. G. A. Somorjai and A. L. Marsh, *Philos. Trans. R. Soc. London, Ser. A*, 2005, **363**, 879.
8. J. A. Jensen, K. B. Rider, M. Salmeron and G. A. Somorjai, *Phys. Rev. Lett.*, 1998, **80**, 1228.
9. E. K. Vestergaard, P. Thostrup, T. An, E. Laesgaard, I. Stensgaard, B. Hammer and F. Besenbacher, *Phys. Rev.*, 2002, **88**, 259601.
10. S. R. Longwitz, J. Schnadt, E. K. Vestergaard, R. T. Vang, E. Laesgaard, I. Stensgaard, H. Brune and F. Besenbacher, *J. Phys. Chem.*, 2004, **108**, 14497.
11. M. Montano, M. Salmeron and G. A. Somorjai, *Surf. Sci.*, 2006, **600**, 1809.
12. A. F. Carley, P. R. Davies, R. V. Jones, K. R. Harikumar, G. U. Kulkarni and M. W. Roberts, *Chem. Commun.*, 2000, 185.
13. P. Thostrup, E. K. Vestergaard, T. An, E. Laegsgaard and F. Besenbacher, *J. Chem. Phys.*, 2003, **118**, 3724.
14. R. W. Joyner and M. W. Roberts, *Surf. Sci.*, 1979, **87**, 501.
15. A. F. Carley, A. Chambers, P. R. Davies, C. G. Mariotti, R. Kurian and M. W. Roberts, *Faraday Discuss.*, 1996, **105**, 225.
16. P. Thostrup, E. Christofferson, H. T. Lorensen, K. W. Jacobsen, F. Besenbacher and J. K. Nørskov, *Phys. Rev. Lett.*, 2001, **87**, 126102.
17. M. Montano, K. Bratlie, M. Salmeron and G. A. Somorjai, *J. Am. Chem. Soc.*, 2006, **128**, 13229.

Further Reading

R. R. Vang, E. Laesgaard and F. Besenbacher, Bridging the pressure gap in model systems for heterogeneous catalysis with high-pressure STM, *Phys. Chem. Chem. Phys.*, to be published.

CHAPTER 8
Molecular and Dissociated States of Molecules: Biphasic Systems

"Science clears the fields on which technology can build"

Werner Heisenberg

8.1 Introduction

Early views on the chemisorption of diatomic molecules and rationalising the specificity that metals exhibited in bond cleavage (dissociative chemisorption) were based on the d-band theory or holes in the d-band concept.[1] Transition metals were active in dissociative chemisorption whereas sp-metals, such as zinc and silver, favoured molecular adsorption. For carbon monoxide, the classical model involved back-bonding of metal electrons into the antibonding orbitals of the adsorbed CO, resulting in bond weakening and cleavage. This was the basis of the Dewar–Chatt or Blyholder models,[2] with the energetics being explained in terms of the energy gained by forming two strong surface bonds. In the case of CO, metal–carbon and metal–oxygen bonds were formed with two adjacent metal sites being a prerequisite for dissociation to take place.

The Lennard-Jones potential energy diagram provided a qualitatively satisfying model, a molecule approaching the surface becoming attracted by a relatively small energy minimum which represents the molecular (or precursor) state followed by a transition over an energy barrier to the dissociated state. Generally for diatomics at transition metals this barrier is small and can be overcome at low temperatures and characterised by a high sticking probability.

Dissociative chemisorption was considered to be either *direct*, when the incoming diatomic molecule has sufficient energy to surmount the barrier without being trapped into the molecular state, or *indirect*, when it passes via the molecular (precursor) state into the dissociated state. If the dissociated state is not immediately equilibrated with the lattice, the fragments will move across the surface, losing energy as they go until equilibrium is finally attained. In 1963, Ehrlich estimated for the dissociative chemisorption of nitrogen at tungsten that up to 50 diffusive hops may be required for equilibrium.[3] The possible relevance of "hopping fragments" in surface-catalysed reactions at single-crystal surfaces became more evident recently from coadsorption studies (see Chapter 2). A detailed treatment of the kinetics involved in chemisorption at metals is considered in many textbooks (see Further Reading), with surface diffusion, the role of lateral interactions and Kisliuk's model for incorporating the precursor state in the model considered. How, then, were these to be viewed with the availability of STM?

It is frequently asserted that two weaknesses of STM are first that all atomic asperities in images need not necessarily correspond to atom surface positions and second that it is inherently difficult to establish the identity of imaged atoms when two or more surface species are involved. The latter need not, however, be a problem. In a study (for example) of the oxidation of ammonia at Cu(110) the oxygen and nitrogen adatoms form separate individual structures which run in the <100> and <110> directions, respectively, whereas under ammonia-rich conditions only imide species are formed, running in the <110> direction, with *in situ* XPS confirming their presence and the absence of surface oxygen (Chapter 5).

8.2 Nitric Oxide

Interest in the chemisorption and surface reactivity of nitric oxide was stimulated initially by the need to understand the chemistry of its catalytic removal from automobile exhausts and subsequently its role in atmospheric chemistry. A review by Shelef[4] in 1975 was dominated by infrared studies, particularly those of Terenin's group in the USSR, with evidence for single- and bridge-bonded NO through comparisons made with nitrosyl complexes. Even though exchange reactions on isotopically labelled nitric oxide and oxide surfaces occurred at high temperatures, N–O bond breaking was not at the forefront of the dynamics proposed for NO chemisorption at metal surfaces in view of the extensive infrared evidence for molecular, nitrosyl-like, species. That NO dissociation might precede the formation of molecular states would not have been recognised by vibrational spectroscopy in the 1960s.

With the development of photoelectron spectroscopy (XPS and UPS) and its ability to distinguish between bonding states, it was established in 1976 that dissociative chemisorption of nitric oxide was facile at some metal surfaces, occurring below room temperature.[5] This had already been established for carbon monoxide at iron surfaces[6] and in view of the smaller bond energy of NO (600 kJ) compared with CO (1000 kJ) it was not surprising. A further point of interest was the observation of N_2O being formed at 80 K with Cu(111) and Cu(100) surfaces and which desorbed at 110 K. This was unexpected and indicative of a "complex chemisorption" process occurring at cryogenic temperatures. The following reaction scheme was suggested[7] based on N(1s) and

O(1s) spectra, which could be assigned to N(a), $N_2O(a)$ and NO(a) and their concentrations calculated from the intensities of the relevant spectra.

$$2NO(g) \rightarrow N(s) + O(a) + NO(a): 80 \text{ K}$$

$$N(s) + NO(a) \rightarrow N_2O(a): 80 \text{ K}$$

$$N_2O(a) \rightarrow N_2O(g): 110 \text{ K}$$

The formation of N_2O was suggested to involve an addition reaction between mobile nitrogen adatoms N(s) and molecularly adsorbed nitric oxide NO(a); two molecular states of NO(a) were also present at 80 K, assigned to bridge and linear states. At 295 K, chemisorption was dissociative, resulting in just chemisorbed oxygen and nitrogen adatoms. The notations (s) and (a) represent transient and the final chemisorbed states, respectively. An alternative view that was later proposed for N_2O formation at cryogenic temperatures was that it involved a dimer mechanism;[8] this was based on vibrational spectroscopic studies.

$$NO(g) \rightarrow (NO)_2(a)$$

$$(NO)_2(a) \rightarrow N_2O(a) + O(a)$$

On the other hand, Kim et al.[9] were of the view, also based on vibrational studies, that at Cu(100) dimer formation could only occur at below 60 K, which is well below that in XPS studies (80 K) when N_2O formation was first reported. What, then, has been learnt from STM?

With Cu(110) at 295 K there are present[10] in the STM images (Figure 8.1) the characteristic $(2 \times 1)O$ rows associated with chemisorbed oxygen running in the <100> direction together with, but well separated from them and running in the <110> direction, evidence for the development of Cu–N chains. The rest of the surface is disordered. The O(1s) and N(1s) spectra with peaks at 530 and 397 eV, respectively, provide confirmation that dissociation of nitric oxide has occurred. On heating to 330, 410 and 430 K, with STM images taken at these temperatures, ordering occurs within the disordered regions to form discrete $(2 \times 3)N$ structures and a biphasic surface structure is evident. The $(2 \times 3)N$ phase is identical with that observed when complete oxydehydrogenation of ammonia is observed at high temperatures.

It is clear that following NO dissociation, the formation of the $(2 \times 1)O$ structure involves facile oxygen mobility; the formation of the well-formed $(2 \times 3)N$ structure is more restricted due to the less mobile nitrogen adatoms, however, and with increasing temperature ordering occurs. Associated with the development of both structures is the diffusion of copper atoms from surface steps to form the new structures.

$$NO(g) \rightarrow O(s) + N(s)$$

$$O(s) \rightarrow (2 \times 1)O$$

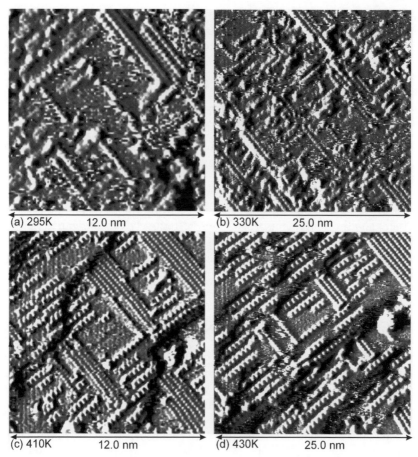

Figure 8.1 STM images of a Cu(110) surface (a) after exposure (25 L) to nitric oxide at 295 K; (b), (c) and (d) after heating (a) to 330, 410 and 430 K, respectively, with the images recorded at the temperatures stated. Note the biphasic structure with nitrogen and oxygen states running at right-angles to each other. (Reproduced from Ref. 10).

$$N(s) \rightarrow \text{disordered } N(a)$$

$$N(a) \rightarrow \text{thermally induced ordering } (2 \times 3)N$$

There have been no STM studies at low temperatures, but what is unequivocal is the observed mobility of the nitrogen and oxygen adatoms, an essential part of one of the mechanisms proposed[7] for the formation of N_2O. Both fragments (nitrogen and oxygen adatoms) have associated with them significant translational kinetic energy attributable to the partitioning of the energy associated with the exothermicity of the bond dissociation and chemisorption process. This concept was first proposed when NO was coadsorbed with ammonia (as a mixture) at Mg(0001), the transient mobile oxygen adatoms, designated O^-, being the active oxidant.[11]

Takehiro et al.[12] have also studied this system (STM only) with similar observations; the –Cu–O–Cu– added row structure and nitrogen features, which initially nucleate near steps, but subsequently are mobile and transform into the (2 × 3)N phase (Figure 8.2). Heating to 370 K increased the ordering of both phases with some loss of nitrogen. The results of both the Aarhus and Cardiff groups are also in general agreement with those reported for nitrogen (atom) adsorption.

Nitric oxide is dissociatively chemisorbed at Ru(0001) at 295 K, with Zambelli et al.[13] establishing the role of a surface step in the dynamics of the dissociation process. Figure 8.3 shows an STM image taken 30 min after exposure of the ruthenium surface to nitric oxide at 315 K. There is clearly a preponderance of dark features concentrated around the atomic step (black strip), which are disordered nitrogen adatoms, while the islands of black "dots" further away

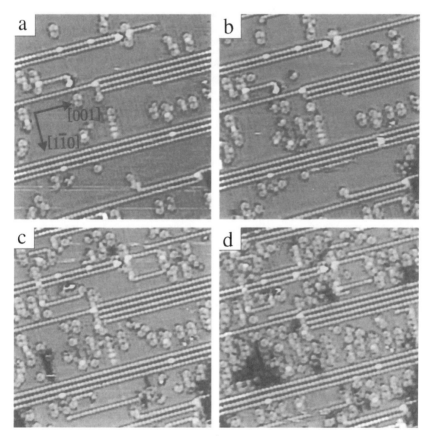

Figure 8.2 Four STM images (175 × 185 Å) recorded over the same area of a Cu(110) surface during the initial exposure to NO at room temperature. There are present the (2 × 1)O strings and isolated features which on heating (see Figure 8.1) agglomerate to form the nitrogen chains running in the <1$\bar{1}$0> direction. (Reproduced from Ref. 12).

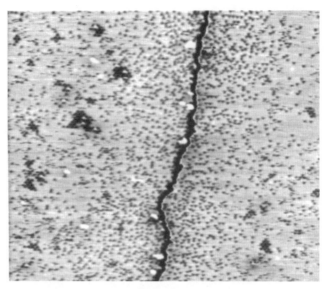

Figure 8.3 STM image (380 × 330 Å) of a Ru(0001) surface with a step after exposure to NO at 315 K; the lower terrace is to the right of the step. The disordered "dots" are N adatoms; the islands consist of oxygen adatoms. (Reproduced from Ref. 13).

from the step are "ordered" oxygen islands. The uneven distribution of the nitrogen and oxygen adatoms, with preferable ordering of the oxygens, is analogous to what is observed with Cu(110); oxygen adatoms are highly mobile, nitrogen adatoms less so. The concentration gradient at the step suggests that dissociation of the nitric oxide molecule occurs at the step, but since both sides of the step were covered with nitrogen atoms the dissociation must have occurred at the upper edge of the step, as the atoms would have been unlikely to diffuse "uphill" over the step. There is no evidence for molecular adsorption of nitric oxide, which, as "the precursor state" is highly mobile, diffuses to the step-edge where it dissociates.

At a Pd(111) surface at room temperature, the chemisorption state is disordered when the NO pressure is less than 3×10^{-6} Torr with very noisy STM images due to the high mobility of the adsorbed molecules.[14] With increasing pressure (and coverage), the c(4 × 2) state, which is reversible, is locked-in and immobile. The adsorption at lower temperatures (150–200 K), where the coverage exceeds that at room temperature, the c(4 × 2) state coexists with a p(2 × 2) and a c(8 × 2) phase; the latter is only present when it coexists with the c(4 × 2) and p(2 × 2) states.

Freund's group at the Fritz Haber Institute have put much emphasis on linking surface science studies with applied catalysts through replicating the latter with model systems without having to resort to the complexity of the real system. A system they have studied in detail is that of nitric oxide chemisorption at a palladium–alumina model catalyst, where they isolated different adsorption and reaction sites on nanoparticles supported on alumina.[15] They made use of the molecular beam approach (Figure 8.4), which enables quantitative kinetic data to be obtained under carefully controlled experimental conditions and thereby provides an insight into molecular events at the microscopic level. Of particular significance in their investigations was the role that atomic fragments, nitrogen and oxygen, could have on both the adsorption and dissociation of nitric oxide using time-resolved infrared reflection absorption spectroscopy.

They concluded that preferential adsorption of nitrogen and oxygen adatoms occurs in the vicinity of edge and step sites on the palladium particles. In other words, NO dissociation is found to be dominated by particle edges, steps, defects and (100) sites rather than by the majority of the (111) facets. Above 300 K, the atomic fragments migrate on to the particle facets. The presence of strongly chemisorbed nitrogen adatoms surprisingly enhanced the probability of NO dissociation.

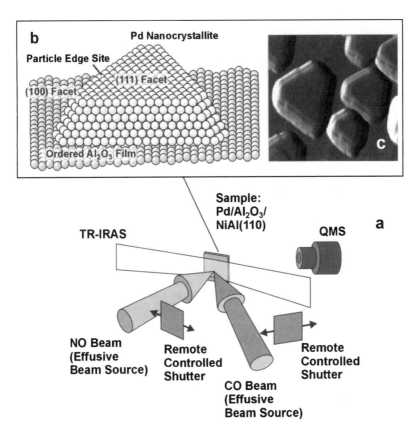

Figure 8.4 (a) Schematic representation of the molecular beam set-up; (b) model of the palladium particles present on the surface of the model catalyst; (c) STM image of the Pd particles on the Al_2O_3–NiAl(110) surface (200 × 200 Å). (Reproduced from Ref. 15).

A key feature of this study was the structural information available on the model palladium nanoparticle catalyst. The mean particle size is 5.5 nm, containing on average 3000 atoms; the majority of the particles are well formed with a (111) orientation and terminated by (111) facets with only a small fraction of (100) facets exposed.

8.3 Nitrogen Adatoms: Surface Structure

Related to the interpretation of the STM studies of nitric oxide dissociation at copper surfaces are the extensive studies during the period 1990–1994 of nitrogen adatoms at Cu(100) and Cu(110) surfaces by a variety of experimental methods: LEED, X-ray scattering, ion scattering, SEXAFS and STM. Nieuhs et al.,[16] using STM, were of the view that at Cu(110) every third <110> row is missing and that images with atomic resolution taken of both ordered (2 × 3) domains and local defective arrangements provided evidence for long-range, highly directional interaction between Cu–N–Cu bonds. The nitrogen adatoms, although adsorbed in each second bridge position along <110>, are aligned in the <100> direction.

Subsequently, Mitchell's group in Vancouver, by means of a tensor-LEED study[17] of the Cu (110)–(2 × 3)N surface structure, supported a reconstruction model in which the topmost layer is described as a pseudo-(100)–c(2 × 2)N overlayer with metal corrugation of about 0.52 Å in the reconstructed layer. Each nitrogen adatom is almost coplanar with the local plane formed by the four neighbouring copper atoms. Of the four N atoms present in the unit mesh, three are also bonded to Cu atoms in the layer below and therefore are five coordinate.

Of crucial significance in deciding between various models have been estimates of the number of copper atoms required to transform the surface into a (2 × 3)N phase. This was the approach adopted by Takehiro et al.[12] in their study of NO dissociation at Cu(110). They concluded that by determining the stoichiometry of the (2 × 3)N phase that there is good evidence for a pseudo-(100) model, where a Cu(1$\bar{1}$0) row penetrates into the surface layer per three [1$\bar{1}$0]Cu surface rows. It is the formation of the five-coordinated N atoms that drives the reconstruction. The authors are of the view that their observations are inconsistent with the added-row model. The structure of the (2 × 3)N phase produced by implantation of nitrogen atoms appears to be identical with that formed by the dissociative chemisorption of nitric oxide.

Nitrogen adatom diffusion is clearly of significance in determining the surface structures observed by STM for NO dissociation at (for example) a Cu(110) surface (Figures 8.1 and 8.2). The Aarhus group has used a combination of fast-scanning STM coupled with *ab initio* DFT calculations to provide a picture of nitrogen adatom diffusion at an Fe(100) surface.[18] The activation energy barrier for the diffusion of isolated N adatoms is (0.92 ± 0.04) eV but is significantly modified when neighbouring adatoms are present. The pre-exponential factor is $4.3 \times 10^{12} \text{s}^{-1}$. There are no comparable data available for copper surfaces.

8.4 Carbon Monoxide

Carbon monoxide has been very much at the heart of industrial and applied catalysis over the last 50 years or more, particularly with its involvement in both Fischer–Tropsch and methanol synthesis. It also has played a very significant role in fundamental studies of chemisorption at metal surfaces, stimulated substantially by carbon monoxide's high absorption extinction coefficient in the infrared region of the spectrum. The interplay between industrial relevance and fundamental understanding, the nature of surface bonding and whether it was chemisorbed molecularly or dissociatively were aspects which were central to experimental studies between 1960 and 1980. This led to the Dewar–Chatt or Blyholder models of the chemisorbed state, its relative ease of dissociation on some metals, such as tungsten or iron and how surface modifiers such as sulfur could control or inhibit the process of dissociation.[6] From a study of XP spectra and a knowledge of the heat of CO chemisorption, ΔH, a correlation was shown to exist[6] between the O(1s) binding energy and ΔH, which enabled the state of the adsorbed CO, molecular or dissociative, to be predicted. The O(1s) binding energy providing a measure of the charge on the oxygen atom arising from back-donation from the metal, the greater the charge the greater is the propensity for C–O bond cleavage.

For CO chemisorption at Ni(110), Sprunger et al.[19] at Aarhus University have reported atomically resolved images of CO molecules. However, at sub-monolayer coverage the molecules are unable to be directly imaged due to their hindered translational and rotational freedom. Tip effects are ruled out. However, at saturation coverage the (2×1)–2CO structure is locked-in and immobile, revealing the well-defined zig-zag structure of the overlayer (Figure 8.5). When coadsorbed with sulfur or oxygen(atoms), well-defined images are observed which indicate a short-bridge geometry of chemisorbed carbon monoxide. There is also evidence for tilting of the CO molecules of about 17° with respect to the surface normal. The influence of coadsorbed oxygen in controlling order–disorder phenomena has also been observed for the Cu(110)–S system.

The $c(4 \times 2)$–2CO structure observed[20] at Ni(111) at room temperature has CO occupying both fcc and hcp threefold hollow adsorption sites with a surface coverage of 0.5 ML. So as to maximise the O–O distance, the molecular axis is tilted away from the surface normal towards atop positions. Corrugation of the adlayer is attributed to a CO-induced buckling of the surface nickel atoms, which is manifested by height differences between adjacent CO molecules (Figure 8.6).

There have been relatively few examples where it has been feasible to determine from an STM image of a chemisorbed adlayer the structural assignment of the adsorption site. This has been due to the difficulty of simultaneously resolving the underlying metal atom substrate. However, Pederson et al.,[21] using the $c(4 \times 2)$CO structure present at Pt(111), have determined the adsorption site by coupling theoretical and experimental images. They established that by placing CO in different adsorption sites only the

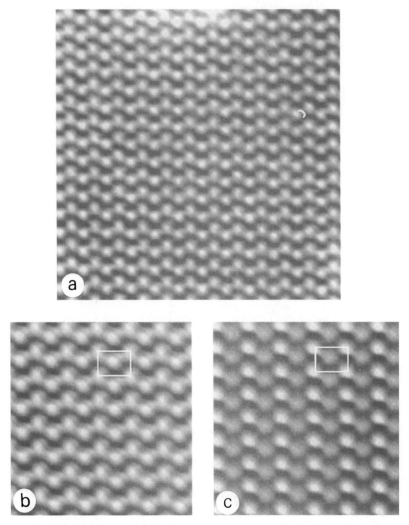

Figure 8.5 STM image of Ni (110) exposed to CO at 1×10^{-6} mbar: (a) raw data (60×60 Å); (b) and (c) unit cell averaged (30×30 Å) at two different tunnelling conditions; the unit cell is indicated. (Reproduced from Ref. 19).

structure with CO in atop bridge sites agree with the experimental images at 300 K. This is in agreement with the earlier assignment, based on vibrational spectroscopy (electron energy loss), by Hopster and Ibach.[22]

Bradshaw's group,[23] by using a very stable high-resolution STM, imaged single molecules of carbon monoxide at Cu(110) at 4 K. At this low temperature, the problems arising from imaging mobile surface species are minimised. This is the only example reported for a single molecule of CO, so that by simultaneously monitoring the corrugation of the substrate copper atoms under conditions of weak tip–surface interaction, it was established that the

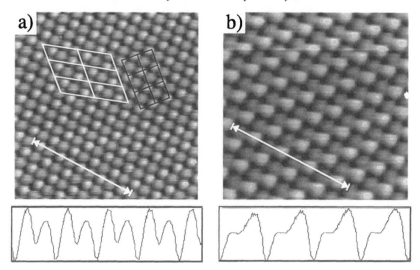

Figure 8.6 STM image of Ni (111)–c(4 × 2)–"CO structure with (a) (4 × 2) (white) and c(4 × 2) (black) unit cells shown with corresponding corrugation line scan (0.2 Å full scale); (b) similar to (a) under different tunnelling conditions and corresponding line scan (0.3 Å full scale). (Reproduced from Ref. 20).

CO occupies the atop site. The authors argue that adsorption, even at 4 K, is in the chemisorbed state with the molecular axis oriented perpendicular to the surface. In a physisorbed state, variations in the orientation, including where the C–O axis is parallel to the surface, would be expected to maximise the van der Waals interaction. The oxidation of CO at Cu(110) is discussed elsewhere (Chapter 5).

8.5 Hydrogen

The classical and traditional view is that for dissociative chemisorption of diatomic molecules to occur at metal surfaces, it is essential that two adjacent (vacant) sites are available:

$$2M + H_2(g) \rightarrow 2M - H(a) \qquad (1)$$

It was an approach that enabled bond energies of chemisorbed states D_{M-H} to be estimated [eqn (2)] provided that the heat of chemisorption ΔH was known, with D_{H_2} the H_2 bond energy:

$$D_{M-H} = \frac{1}{2}(D_{H_2} + \Delta H) \qquad (2)$$

This model has, however, never been established experimentally and one can envisage that, as was suggested for the dissociative chemisorption of dioxygen, the process could involve just one surface site with the exothermicity associated with the formation of a single M–H bond resulting in the second hydrogen atom having translational kinetic energy sufficient for it to undergo diffusion (surface hopping) over a number of surface sites (see Chapters 2 and 4). Equation (2) would then be invalid. STM confirmed what for oxygen was termed "abstractive chemisorption" and a model proposed for explaining oxygen reactivity patterns in coadsorption studies.

Salmeron, in an elegant study,[24] applied STM to explore whether the "dual site" hypothesis was applicable for hydrogen dissociation at Pd(111), resorting to cryogenic studies to slow down the chemisorption process. He considers first the fact that at temperatures below 50 K, where the hydrogen atom coverage approaches 0.66, further adsorption of H atoms ceases and the surface remains in the $\sqrt{3} \times \sqrt{3}$ R30°–2H structure, which covers most of the surface. In this case each H vacancy is separated from other vacancies by H adatoms, which suggests that no further dissociative chemisorption can occur at isolated or single vacant sites. The coverage, however, can be increased to approach unity by increasing the temperature above 50 K in the presence of $H_2(g)$, resulting in surface diffusion setting in and leading to vacancy aggregation, *i.e.* the creation of vacancy ensembles involving two, three or four vacant sites. Salmeron is of the view that it is the observation of "vacant site" aggregation accompanying the increase in hydrogen adatom coverage that is the experimental evidence that is central to the mechanism of the dissociative chemisorption of hydrogen (Figure 8.7).

What is also evident is that a *single vacancy site* is not active in the dissociation process – for the abstractive chemisorption process in this case, one of the hydrogen atoms could have undergone surface diffusion and seek out a second vacant site as with dioxygen at (for example) an Al(111) surface. On increasing the temperature above 50 K, the $\sqrt{3} \times \sqrt{3}$ R30°–2H structure becomes disordered with facile surface "vacant site" movement leading to the possibility of dimers, trimers and tetramers. STM images taken at 65 K show diffusing vacancies at close to a monolayer coverage of H adatoms. The images show the formation of aggregates of two vacancies. The dimer vacancy has the appearance of a three-lobed object due to the rapid diffusion of a H atom next to the dimer. This H atom can undergo "hopping" over a bridge site (low energy barrier) to occupy a vacancy site. When the dimer breaks up (dissociates), two isolated vacant sites are then observed.

Triplets of vacant sites were also observed, although at a slower rate than for the dimers; these, in contrast to the dimers, were active in the dissociative chemisorption of hydrogen. STM images show in a movie the precise moment of the formation of two vacancy triplets. The three initially isolated vacant sites form two trimer images as a bright triangle. At 65 K, in vacuum, the lifetime of triplets before breaking down into vacancies or dimers is 12 min. However, in the presence of $H_2(g)$ at a pressure of 2×10^{-7} Torr, the triplets were filled up by two H adatoms from a dissociatively adsorbed hydrogen molecule, which transformed the triplet into a single vacancy. Active sites for H_2 dissociation were also formed by ensembles of four or more vacancies, but these were observed less frequently than dimers or trimers.

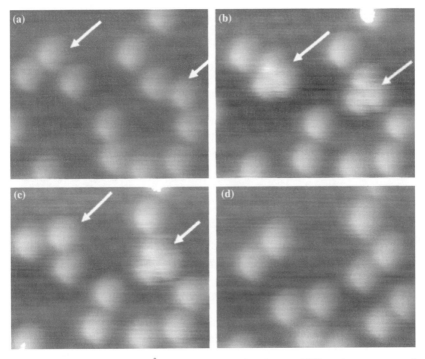

Figure 8.7 Frames (23 by 35 Å) of an STM movie taken at 65 K at close to a complete monolayer of hydrogen adatoms at Pd(111) showing vacancy diffusion. The images (b) and (c) show the aggregation of two nearest neighbour vacancies, which has the appearance of a three lobed object due to the rapid diffusion of one H atom next to the vacancy dimer. (Reproduced from Ref. 24).

The general conclusion is that at least three neighbouring vacancies are necessary to dissociate hydrogen, with density functional theory indicating that the dissociation probability is some 10^6 times greater at a trivacancy "open structure", in agreement with the experimentally observed inactivity of single and dimer vacancy sites. This is a conclusion with far-reaching implications.

8.6 Dissociative Chemisorption of HCl at Cu(110)

When a Cu(110) surface is exposed to hydrogen chloride at 295 K, the chlorine adatoms are initially mobile and disordered but with time a c(2 × 2) structure forms. (Figure 8.8a). From the intensities of the Cl(2p) and Cu(2p) spectra, the concentration of chlorine atoms in calculated (see Chapter 2) to be $5.1 \times 10^{14}\,\mathrm{cm}^{-2}$, very close to what would correspond to a monolayer. However, when the HCl exposure is increased to about 500 L, the surface chlorine adatom concentration is increased to $6.1 \times 10^{14}\,\mathrm{cm}^{-2}$. The STM image (Figure 8.8b) indicates that buckling of the surface occurs in the [001] direction, with the buckled rows 18 Å apart. Accompanying the formation of the c(2 × 2)Cl

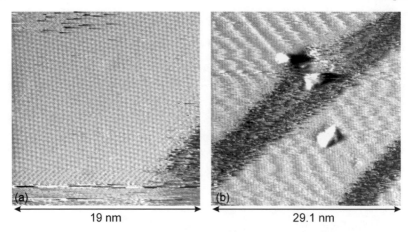

Figure 8.8 (a) c(2 × 2)Cl adlayer ($\sigma = 5.1 \times 10^{14}$ Cl adatoms cm^{-2}) at Cu(110) formed by dissociation of hydrogen chloride at 295 K; (b) buckling of c(2 × 2) Cl adlayer at 295 K, an example of "corrosive chemisorption". (Reproduced from Ref. 35).

adlayer, there is also appreciable mobility of copper adatoms including step movement (see Chapter 4) and surface buckling. That halogen adatoms increased the surface self-diffusion of copper substrate atoms was reported by Delamare and Rhead[33] in 1971 with Walter et al.[34] in 1996 reporting a LEED–AES study of the Cu(111)–Cl system. Both studies concluded that "a mixed adsorbent–adsorbate layer" is formed akin to "a quasi-two-dimensional liquid in which diffusion is rapid".

8.7 Chlorobenzene

The manipulation of individual atoms and molecules with the STM was one of the outstanding discoveries during its early development. Eigler and his colleagues demonstrated[25] how to assemble predesigned nanometre-scale structures which could trap electrons. For those interested in molecular events at surfaces related specifically to the mechanism of chemisorption and catalytic reactions, it was essential to rule out the influence of the "tip" in an STM study, and this has been an aspect that investigators took great care to rule out as a factor in their experiments.

Palmer's group at Birmingham investigated[26] the dynamics of STM-induced desorption and dissociation of chlorobenzene molecules at Si(111)–(7 × 7). Figure 8.9 shows images of adsorbed chlorobenzene before and after desorption. The authors were particularly interested in why energy can be channelled so precisely to break specific chemical bonds, particularly since the C–Cl bond is 3.6 eV and the surface adsorption binding energy is about 1.0 eV. They established that the manipulation mechanism depended crucially on the current and voltage between the tip and the surface. The desorption rate varies linearly

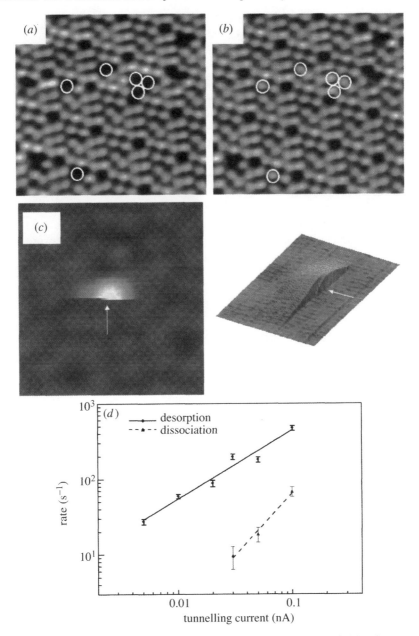

Figure 8.9 Dynamics of STM-driven desorption and dissociation of chlorobenzene at Si(111)–(7 × 7) (a) before and (b) after a desorption scan; the circles indicate the positions of chlorobenzene molecules before and after desorption; (c) appearance of a chlorine adatom formed by dissociation of chlorobenzene with corresponding 3D image; (d) measured rates of desorption and dissociation as a function of tunnelling current for a sample bias of +3 V. (Reproduced from Ref. 26).

with the current (at a fixed voltage) and is essentially independent of the tip–surface distance, i.e. independent of the electric field, ruling out an electric field mechanism. The desorption yield is, however, dependent on the sample bias and they concluded that the desorption mechanism is current driven.

The second event in STM manipulation of the chlorobenzene system is *dissociation* of the C–Cl bond and the fate of the chlorine atom. Figure 8.9 shows the measured rates of molecular dissociation and desorption as a function of tunnelling current at a fixed voltage (+3 V). The desorption rate is linearly dependent on current (0.9 ± 0.1), indicating that it is controlled by a single tunnelling electron through the attachment of an electron to the antibonding π^* state of the adsorbed chlorobenzene resulting in vibrational excitation of the molecule-surface bond. In contrast, dissociation is a two-electron process, the current dependence of the dissociation rate being 1.8 ± 0.3. The authors explain this through a mechanism that couples vibrational excitation and dissociative electron attachment steps. The experiments were based on STM images obtained at room temperature where the dissociation of a single chlorobenzene molecule was followed under different voltage or current conditions.

8.8 Hydrocarbon Dissociation: Carbide Formation

The group at Aarhus have reported carbon-induced structures at Ni(111) and Ni(110) surfaces resulting from the dissociation of ethylene at high temperatures.[27] Between 400 and 500 K, the Ni(110) surface is seen to form two carbidic structures with (4×3) and (4×5) domains present arising from surface reconstruction with substantial transport of nickel taking place. At higher temperatures (560 K), the surface becomes dominated by the (4×5) structure, which is well ordered and can be observed clearly by LEED. Ion scattering studies provide additional information which enables models to be constructed for both the (4×3) and (4×5) phases.

The Ni(111) surface reconstructs when exposed to ethylene at 500 K to form an almost square, $(5 \times 5)\,\text{Å}^2$, (2×2)–2C surface mesh. The carbon atoms thereby increase their coordination to the nickel atoms, which is the driving force for the reconstruction.

8.9 Dissociative Chemisorption of Phenyl Iodide

Phenyl iodide chemisorbs dissociatively at a Cu(110) surface at 295 K with structural information being obtained from STM and chemical information from XPS.[28] At low exposures (6 L), the surface concentrations of carbon and iodine species, calculated from the intensities of the C(1s) and I(3d) spectra, were in the expected 6:1 ratio and the iodine concentration $5.1 \times 10^{14}\,\text{cm}^{-2}$. With further exposure, the iodine concentration increased and reached a maximum value of $5.5 \times 10^{14}\,\text{cm}^{-2}$ after an exposure of 1200 L. This was

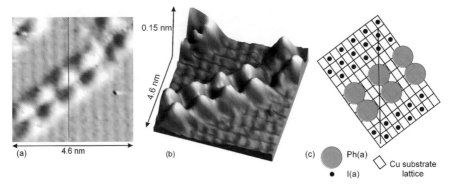

Figure 8.10 STM images showing coexisting c(2 × 2)I(a) and phenyl adsorbates at a Cu(110) surface. (a) After 180 L PhI, $V_S = -2.88$ V, $I_T = 1.41$ nA; note the offset between the maxima in the iodine lattice either side of the phenyl chain showing that the phenyl groups are situated in a grain boundary in the iodine lattice. (b) 3D representation of (a) showing clearly the I(a) maxima. (c) Schematic model of the coexisting iodine and phenyl lattices. (Reproduced from Ref. 28).

accompanied by the loss of "carbon" through desorption of the phenyl groups through a coupling reaction to form biphenyl. The chemisorbed iodine is present in a c(2 × 2) structure (Figure 8.10) with the phenyl groups appearing as roughly circular images, 0.7 nm in diameter and 0.1 nm above the plane of the iodine adlayer. The phenyl groups preferentially present at step-edges resist desorption and extend for up to 20 nm in the <110> direction. Chains of phenyl groups, stabilised by pair formation in the <100> direction, are also observed on the terraces (Figure 8.10).

8.10 Chemisorption and Trimerisation of Acetylene at Pd(111)

That cyclomerisation of acetylene to form benzene at Pd(111) was first reported in 1983 by a number of groups[29] from spectroscopic studies. With most other metals, acetylene undergoes dissociation with carbon–carbon bond cleavage. In a detailed study in 1998, Janssens *et al.*,[30] using STM, explored the conditions under which trimerisation occurs over the temperature range 140–230 K. Initially a well-ordered c(2 × 2) overlayer forms, but with increasing coverage at 140 K it is compressed to a (3 × 3) $R30°$ overlayer and this also occurs when the saturated overlayer is warmed to 230 K (Figure 8.11).

Cyclotrimerisation at 140 K takes place almost exclusively in the *disordered areas* between the domains of (3 × 3) $R30°$ but simultaneously with the formation of this phase. Benzene formation stops when the (3 × 3) $R30°$ phase is close to completely covering the surface. No further reaction occurs on further exposure to acetylene or on heating to 200 K, establishing that the acetylene

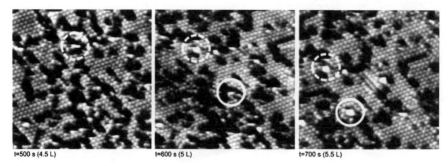

Figure 8.11 Sequence of STM frames of acetylene on Pd(111) at 140 K (150 × 150 Å, 0.8 nA, 91 mV). Scanning rate: 100 s/frame. Exposure times and approximate doses as indicated. The sequence shows the formation of benzene and the further saturation of the (3 × 3) $R30°$ layer. The circles mark the appearance of bright features attributed to benzene molecules. (Reproduced from Ref. 30).

molecules present in this phase are not the reactive species involved in the formation of benzene. The reactive species have not been identified by either STM or by surface-sensitive spectroscopies and are suggested to be a *transient adsorption state* present as a minority species, formed during the compression of the initial (2 × 2) phase to the (3 × 3) $R30°$ configuration and essential for cylcotrimerisation to occur. Prior to these STM studies, at the Fritz Haber Institute there was the view that the reactive acetylene (to form benzene) was that associated with the predominant phase observed by LEED – the (3 × 3) $R30°$ structure.

8.11 Summary

STM has revealed unique information on the dissociative chemisorption of molecules at metal surfaces, a feature being the structural complexity leading to biphasic systems, with surface mobility of both adsorbates and substrate atoms involved. Surface steps and vacancies play a role both in the dissociation event leading to bond cleavage and in providing sites where a fragment is preferentially adsorbed. Salmeron's study of hydrogen dissociation at Pd(111) draws attention to the special significance of vacancies – particularly when aggregated. At least three neighbouring vacancies are claimed to be necessary for the dissociation of hydrogen, with images obtained at cryogenic temperatures essential to reveal their influence in "real time". The model is unprecedented and its more general implications need to be explored.

Phenyl iodide dissociates at Cu(110) to form a c(2 × 2) iodine layer, accompanied by coupling of phenyl groups which desorb as biphenyl but with evidence that some phenyl groups remain at the surface stabilised as chains at step-edges and on terraces as "paired chains". Chemisorption of HCl at Cu(110) is "corrosive", with evidence for surface buckling.

Trimerisation of acetylene at Pd(111) is suggested from STM studies to involve an acetylene transient present in a disordered part of the surface between well-ordered (3 × 3) $R30°$ patches. There is no spectroscopic evidence (due to its low concentration) for the transient and the STM evidence is based on carefully designed experiments, but is essentially circumstantial. It is worth drawing attention, however, to the growing evidence that is emerging for the role that surface transient states, structurally disordered and mobile, can have in surface catalysis and by implication the possible limitations of classical experimental and theoretical approaches. Somorjai and Marsh's studies of ethylene hydrogenation,[31] Wang and Barteau's of the oxidation of butane to maleic anhydride,[32] the trimerisation of acetylene[30] and the evidence for oxygen transients in controlling reaction pathways in catalytic oxidation chemistry (Chapter 5) are further examples. How immeasurably low concentrations of surface transients can provide efficient reaction pathways to products is discussed in Chapter 2. A characteristic feature of nitrogen adatoms is their reluctance to form well-ordered structures, preferring to remain as isolated adatoms unless thermally activated. They are, therefore, strong candidates to participate in an addition reaction to form N_2O accompanying NO dissociation at low temperatures[7] (see also Chau *et al.* in Further Reading).

References

1. D. A. Dowden, *J. Chem. Soc.*, 1950, 242; B. M. W. Trapnell, *Proc. R. Soc. London, Ser. A*, 1953, **218**, 566.
2. G. Blyholder, *J. Phys. Chem.*, 1964, **68**, 2772.
3. G. Ehrlich, *Ann. N. Y. Acad. Sci.*, 1963, **101**, 722.
4. M. Shelef, *Catal. Rev.*, 1975, **11**, 1.
5. K. Kishi and M. W. Roberts, *Proc. R. Soc. London*, 1976, **353**, 289; D. W. Johnson, M. H. Matloob, M. W. Roberts, *J. Chem. Soc., Chem. Commun.*, 1978, 40; *J. Chem. Soc., Faraday Trans. 1*, 1979, **75**, 2143.
6. K. Kishi and M. W. Roberts, *J. Chem. Soc., Faraday Trans. 1*, 1975, **71**, 1715; C. S. McKee and M. W. Roberts, *Chemistry of the Metal–Gas Interface*, Clarenden Press, Oxford, 1979, p. 373.
7. C. T. Au, A. F. Carley and M. W. Roberts, *Philos. Trans. R. Soc. London, Ser. A*, 1986, **318**, 61; M. W. Roberts, *Catal. Lett.*, 2004, **93**, 29.
8. W. A. Brown and D. A. King, *J. Phys. Chem. B*, 2000, **104**, 2578.
9. C. M. Kim, C. -W. Yi and D. W. Goodman, *J. Phys. Chem. B*, 2002, **106**, 7065.
10. A. F. Carley, P. R. Davies, K. R. Harikumar, R. V. Jones, G. U. Kulkarni and M. W. Roberts, *Top. Catal.*, 2001, **14**, 101.
11. C. T. Au and M. W. Roberts, *J. Chem. Soc., Faraday Trans. 1*, 1987, **83**, 2047.
12. N. Takehiro, F. Besenbacher, E. Laegsgaard, K. Tanaka and I. Stensgaard, *Surf. Sci.*, 1998, **387**, 145.

13. T. Zambelli, J. Wintterlin, J. Trost and G. Ertl, *Science*, 1996, **373**, 1688.
14. K. H. Hansen, Z. Sljivancanin, B. Hammer, E. Laesgaard, F. Besenbacher and I. Stensgaard, *Surf. Sci.*, 2002, **498**, 1.
15. V. Johaneck, S. Schauermann, M. Laurin, S. Chinnakonda, S. Gopinath and H. -J. Freund, *J. Phys. Chem. B*, 2004, **108**, 14244.
16. H. Nieuhs, R. Spitzel, K. Besocke and G. Comsa, *Phys. Rev. B*, 1991, **43**, 12619.
17. D. T. Vu and D. A. R. Mitchell, *Phys. Rev. B*, 1994, **49**, 11515.
18. M. Ø. Pedersen, L. Osterlund, J. J. Mortensen, M. Mavrikakis, L. B. Hansen, I. Stensgaard, E. Laesgaard, J. Horsdov and F. Besenbacher, *Phys. Rev. Lett.*, 2000, **84**, 4898.
19. P. Sprunger, F. Besenbacher and I. Stensgaard, *Surf. Sci.*, 1995, **324**, L321.
20. P. T. Sprunger, F. Besenbacher and I. Stensgaard, *Chem. Phys. Lett.*, 1995, **243**, 439.
21. M. Ø. Pedersen, M.-L. Bocquet, P. Sauter, E. Laesgaard, I. Stensgaard and F. Besenbacher, *Chem. Phys. Lett.*, 1999, **299**, 403.
22. H. Hopster and H. Ibach, *Surf. Sci.*, 1978, **77**, 109.
23. M. Doering, J. Buisset, H. -P. Rust, B. G. Briner and A. M. Bradshaw, *Faraday Discuss.*, 1996, **105**, 163.
24. M. Salmeron, *Top. Catal.*, 2005, **36**, 55.
25. M. Crommie, C. P. Lutz and D. M. Eigler, *Science*, 1993, **262**, 218.
26. R. E. Palmer, P. A. Sloan and C. Xirouchaki, *Philos. Trans. R. Soc. London, Ser. A*, 2004, **362**, 1195; P. A. Sloan, M. F. G. Hedouin, R. E. Palmer and M. Persson, *Phys. Rev. Lett.*, 2003, **91**, 118301.
27. C. Klink, I. Stensgaard, F. Besenbacher and E. Laegsgaard, *Surf. Sci.*, 1995, **342**, 250; 1996, **360**, 171.
28. A. F. Carley, M. Coughlin, P. R. Davies, D. J. Morgan and M. W. Roberts, *Surf. Sci.*, 2004, **555**, L138.
29. W. T. Tysoe, G. L. Nyberg and R. M. Lambert, *J. Chem. Soc., Chem. Commun.*, 1983, 623; W. Sesselmann, W. Wonatschek, G. Ertl and J. Kuppers, *Surf. Sci.*, 1983, **130**, 245.
30. T. V. W. Janssens, S. Volkening, T. Zambelli and J. Wintterlin, *J. Phys. Chem. B*, 1998, **102**, 6521.
31. G. A. Somorjai and A. L. Marsh, *Philos. Trans. R. Soc. London, Ser. A*, 2005, **363**, 879.
32. D. X. Wang and M. A. Barteau, *Catal. Lett.*, 2003, **90**, 7.
33. F. Delamare and G. E. Rhead, *Surf. Sci.*, 1971, **28**, 267.
34. W. K. Walter, D. E. Manolopoulos and R. G. Jones, *Surf. Sci.*, 1996, **348**, 115.
35. A. F. Carley, P. R. Davies and M. W. Roberts, *Philos. Trans. R. Soc. London, Ser. A*, 2005, **363**, 829; in *Turning Points in Solid State, Materials and Surface Chemistry*, ed. K. D. M. Harris and P. P. Edwards, Royal Society of Chemistry, Cambridge, to be published.

Further Reading

F. Besenbacher, I. Stensgaard, J. K. Norskov and K. W. Jacobsen, Chemisorption of H, O and S on Ni(110): general trends, *Surf. Sci.*, 1992, **272**, 334.

G. A. Somorjai and M. Yang, The surface science of catalytic selectivity, *Top. Catal.*, 2003, **24**, 61.

T. -D. Chau, T. Visart de Bocarmé and N. Kruse, Formation of N_2O and $(NO)_2$ during adsorption on Au 3D crystals, *Catal. Lett.*, 2004, **98**, 85.

CHAPTER 9
Nanoparticles and Chemical Reactivity

"The electron is not as simple as it looks"

William Lawrence Bragg

9.1 Introduction

Although much of the understanding of fundamental aspects of heterogeneous catalysis has emerged from studies of single-crystal metal surfaces, industrial catalysts are complex systems frequently of small particles supported on oxides such as alumina, nickel oxide, titanium dioxide and silica. The size and shape of the metal particles are generally considered to play a significant role in determining both reactivity and selectivity patterns. When the particles are in the nano-range with clusters consisting of no more than a few atoms, then quantum size effects are often suggested to play a significant role. There are two questions to be addressed: (a) the growth mechanism of the metal particles – is there a critical size or shape prerequisite for exhibiting catalytic activity? and (b) the fundamental reason for the chemistry associated with the observed catalytic behaviour. It is also important that we have a good understanding of the oxide support on which the particles are deposited. The support can be either cleaved or formed *in situ* by oxidation of the metal. Surfaces such as alumina and titanium dioxide are difficult to prepare by cleavage, whereas *in situ* oxidation of metals such as titanium, copper and nickel to form thin oxide films offers certain advantages for investigation by surface spectroscopies but requires careful study to define the surface at the atomic level, particularly whether they are "perfect" or exhibit defects. The "nickel oxide" overlayer at a nickel surface is a good example of where care is necessary, with XPS evidence for Ni^{3+} and O^- states being present when thin oxide overlayers are present at room temperature.[1,2] This is not surprising since, depending on the preparation conditions, "bulk" nickel oxide can be coloured either green or black – the latter being associated with a defective oxide revealed by XPS.[2] Freund reviewed studies of metal deposits on metal oxides taking examples from various research groups where a range of surface sensitive spectroscopies had been used.[3] Goodman's group, for example, deposited nickel on thin films of oxidised aluminium and studied their morphological changes with both increasing coverage and temperature. A feature of this work was also the use of scanning tunnelling spectroscopy (STS), which has not been used extensively in the characterisation of nanoparticles on solid surfaces;[4] a further example is a study of Au clusters at a TiO_2 (110) surface.[5]

Although nanoparticles supported on oxide surfaces are a common feature of industrially relevant catalysts, there have been by comparison with gold particles relatively few STM studies. Other experimental methods based on STEM BF and STEM HAADF have, however, been applied successfully to characterise structural features and these are also considered. There is, however, one area, pioneered by the group at the Royal Institution, London, where advantage has been taken of the ability to design single-site catalysts within open-structure oxidic solids with the detailed environment of these sites determined by EXAFS. In these systems, the well-defined active sites are distributed in a spatially uniform fashion and accessible to the reactants throughout the bulk of the solid. This provides the opportunity to examine by *in situ* methods the relationships that might exist between structure and catalytic effectiveness. There is much emphasis given to the similarities to the approach used by enzymologists – both groups dealing with nanoparticles.

We have not considered the physics of nanoparticles other than when it is relevant to the conditions that control their stability or size and therefore influence the preparation of surfaces relevant to catalysis. Of particular interest is the transition from an insulator to a metallic cluster – at what cluster size does this occur?

9.2 Controlling Cluster Size on Surfaces

Although the control of "gas-phase" clusters was well established in the 1980s, size selection of metal clusters on solid surfaces offers some challenges, with facile surface diffusion being one of them. Palmer and his group in Birmingham used a magnetron sputtering gas aggregation source for the preparation of clusters on solid surfaces; in essence, it consisted of radiofrequency plasma which is ignited in a mixture of helium and argon gas and confined close to a metal target by a magnetic field. Palmer's group investigated initially silver clusters deposited on graphite surfaces at room temperature.[6] Two deposition energy regimes were used where clusters can be immobilised at the surface – no surface diffusion occurring. At high energies (20 eV per cluster atom and greater), the nanoscale clusters can be implanted into the surface and come to rest at the bottom of an open "well". At sufficiently high impact energies (e.g. 2 keV), an array of clusters can be observed, all of similar size, indicating that no cluster aggregation has occurred. Figure 9.1 shows STM and size distributions for Ag^+_{147} clusters deposited at 2 keV on graphite. The narrowness of the

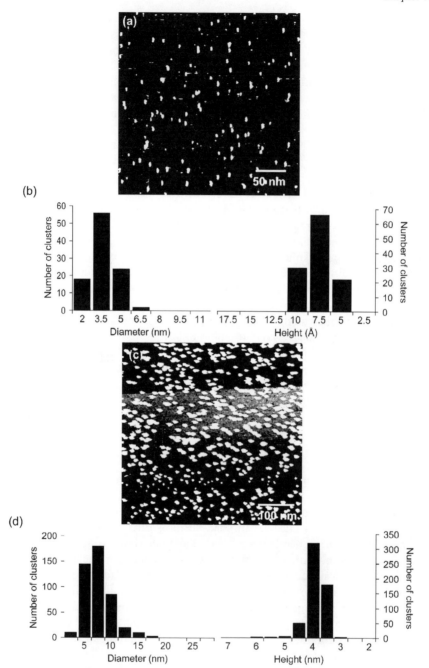

Figure 9.1 (a) STM image of Ag^+_{147} clusters deposited at 2 keV on graphite; (b) the corresponding diameter and height distributions; (c) image of Ag^+_{2700} clusters deposited on graphite at 0.65 keV pretreated by Ar^+ bombardment; (d) the corresponding diameter and height distributions. (Reproduced from Ref. 6).

distribution is consistent with a 147 ± 7 atom silver cluster of height 1–2 monolayers. The pinning of high-energy clusters is one way for preparing well-defined nanoscale surface structures from size-selected clusters with possible applications in catalyst preparation. A second approach is to pretreat the surface with high-energy Ar^+ to induce the formation of surface defects that enable silver clusters to be prepared, which preserve the cluster size generated by the cluster beam source. The defects inhibit the surface diffusion and aggregation of the clusters, an example being $Ag^+{}_N$, where $N = 2700$, prepared at a graphite surface at low energy (650 eV). In the absence of surface defects, high-energy beams (2 keV) would be necessary to immobilise the clusters.[7]

9.3 Alloy Ensembles

Although morphological aspects of gold particles in catalysis have been a major interest recently by STM, one of the earliest studies was that by Sachtler, Biberian and Somorjai[8] in 1981, who, following the work of Sinfelt in the area of catalysis by alloys, investigated gold deposited on Pt(100) and also platinum deposited on Au(100). They used the dehydrogenation of cyclohexene to benzene as the test reaction. Sinfelt had already established the existence of bimetallic clusters with unusual thermodynamic and structural properties and Sachtler et al.[8] showed that the reactivity of Pt(100) was increased by a factor of six by the presence of a monolayer of gold (Figure 9.2). Above a monolayer, the activity decreases. The gold adlayer grows layer by layer (Frank–van der Merwe mechanism) while platinum deposited on Au(100) grows via the formation of microcrystallites (Volmer–Weber mechanism). LEED and AES were used to monitor the growth process. With the development of STM, atom resolved information became available with the possibility of following catalytic activity and correlating it with atomic resolution of nanoscale clusters.

The concept of ensembles associated with the catalytic behaviour of alloy surfaces had been discussed by Kobozev, Dowden, Sinfelt and others.[9] For example, in 1978, van Barneveld and Ponec[10] prepared Ni–Cu non-porous alloy powders of various bulk compositions and studied their selectivity in Fischer–Tropsch synthesis. Activity towards the formation of higher hydrocarbons decreases with increase in the copper content of the alloys and this (for methane formation) is suggested to be due to the decrease in the number of ensembles of nickel which are active in CO dissociation. The latter had been established as a facile process at nickel and iron surfaces by photoelectron spectroscopy.[11] STM, however, provided a means of designing at the atomic level surface alloys with specific properties (Figure 9.3). Through a combination of theory and molecular beam experiments, the Aarhus group,[12] using methane activation as a model system, were able to predict the observed activity of the Au–Ni(111) surface in terms of the special sites created as a function of the gold coverage. The "alloy sites" were found to have a lower activity for methane dissociation than that of the atomically clean Ni(111) surface. It should be emphasised, however, that selectivity is frequently more significant in catalysis than activity. In the

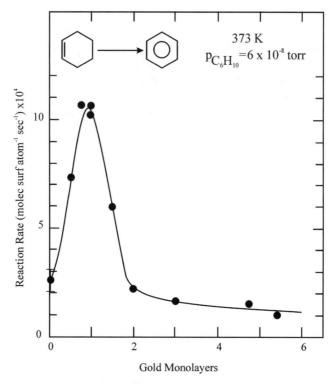

Figure 9.2 Variation of the rate of cyclohexene dehydrogenation to benzene with gold coverage at Pt(100) at 373 K. (Reproduced from Ref. 8).

steam-reforming reaction hydrocarbons (mainly CH_4) and water are converted into hydrogen and carbon monoxide using nickel-based catalysts, but a significant problem has been the simultaneous formation of graphite, which poisons the catalytic reaction. Ensembles of nickel and gold were, however, considered to offer an approach for inhibiting the incorporation of carbon into the nickel catalyst; the industrial approach is to add "sulfur" to the reactants, which selectively poisons carbon formation.

The mechanism of surface alloying was studied by Behm's group,[13] applying STM to investigate the influence of adisland formation for nickel particles deposited on Au(110)–(1 × 2). The authors focused on possible pathways in alloy formation and whether energetic factors or the role of the intermixing process itself is significant. The relevance of the adatom–adatom exchange process (the classical diffusion mechanism) and whether the process can be influenced by the presence of adislands is the thrust of this paper.

9.4 Nanoclusters at Oxide Surfaces

Titanium dioxide is one of the most intensely studied oxides in view of both its use as a support and its special photocatalytic properties. Of special recent

Nanoparticles and Chemical Reactivity

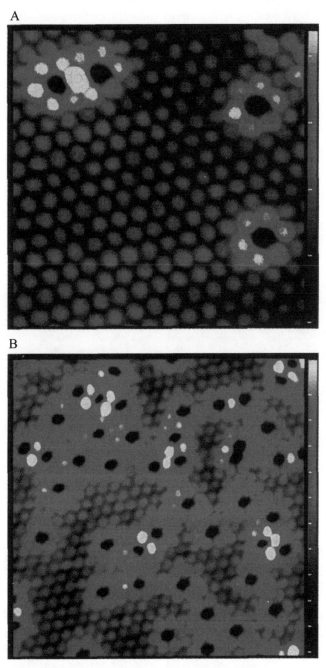

Figure 9.3 Images of an Ni(111) surface (A) with 2% and (B) with 7% of a monolayer of gold. The gold atoms appear black in the images and the nickel atoms adjacent to the gold atoms are brighter (yellow) because of a change in their electronic structure. (Reproduced from Ref. 12).

interest has been its use as a support for gold nanoclusters, shown first by Haruta and subsequently others[14] to be active in low-temperature oxidation catalysis. How the gold particles are attached to the surface, their configuration and the number of gold atoms involved in the cluster were factors considered to control the catalytic activity. Through a combination of STM and theory, the Aarhus group investigated gold clusters on TiO_2, establishing that the growth of gold clusters on the rutile surface can be correlated with the presence of oxygen vacancies[15] (Figure 9.4). Through studies of the temperature dependence of the cluster size distribution and the oxygen vacancy density, the authors establish that oxygen vacancies are the active nucleation sites, that a single Au atom vacancy is stable up to room temperature and that it can bind on average three to five gold atoms. For larger clusters, the gold–oxide interface contains a high density of oxygen vacancies, which increase the binding of gold particles to the oxide surface. The larger clusters formed at room temperature are located at step-edges, presumably because the latter can be considered as an accumulation of oxygen vacancies. Nucleation at step-edges or line defects is now

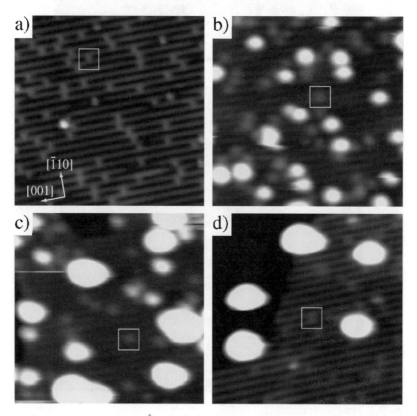

Figure 9.4 Images (150 × 50 Å) of (a) TiO_2(110) surface with bridging oxygen vacancies; (b), (c) and (d) are images of the surface after deposition of 4% ML of Au at 130, 210 and 300 K, respectively. Vacancies are indicated by squares. (Reproduced from Ref. 15).

generally accepted for most nanoparticles, with Ostwald ripening sometimes being observed, as for example by Edgell's group at Oxford[16] for gold particles at TiO_2 [110] at high temperatures.

The author's reported[16] that gold clusters with dimensions of about 50 Å are remarkably static even over a period of 6 h at 750 K. This lack of mobility of medium-sized clusters is suggested to be somewhat unexpected in view of simulation studies by Luedke and Landman,[43] who reported jumps of the order of 200 Å in time intervals as short as 10 ns for Au_{140} clusters on the basal plane of graphite. On the other hand, bimetallic clusters formed when platinum is deposited on rhodium clusters present on TiO_2 are found to be mobile even at room temperature. Shaikhutdinov and his colleagues[17,18] in Berlin studied the details of nanoparticle growth of palladium on FeO(111). At submonolayer coverage, Pd randomly nucleates and forms two-dimensional islands at submonolayer coverage. At 600 K sintering occurs, forming extended two-dimensional islands at low coverage and a thick Pd (111) at high coverage. How these changes influence the chemisorption of carbon monoxide was investigated by TPD, IRAS and molecular beam methods.

Starr et al.[19] studied the formation of gold particles on FeO(111); this work was significant in that the oxide surfaces were free of line and point defects (vacancies) and with wide flat terraces. Therefore, changes observed by STM should be due exclusively to the interaction of the gas with the gold particles. Any changes that might be observed in the terrace structures would provide evidence for the influence of the tip; no changes were observed. Two significant points emerged from this study: (a) that carbon monoxide (99.995% purity!) induced changes in the morphology and stability of the gold particles located at step-edges; the gold particles were stable in oxygen and hydrogen at pressures up to 2 mbar; (b) that impurities present in the 99.995% pure CO could lead to the incorrect conclusion that it was dissociatively chemisorbed at TiO_2–Au surfaces. The latter is a well-known problem in high-pressure "surface science studies"; the authors attributed STM changes observed with unpurified (99.995% pure) CO to traces of metal carbonyls.

Although there have been relatively few STS studies of nanoparticles relevant to catalysis, Goodman's group[5] correlated the onset of catalytic activity of gold clusters on TiO_2 with the development of metallic clusters (Figure 9.5). This involved recording current–voltage curves for a single cluster; the smaller clusters show a behaviour expected of that from a system with a band gap which was absent with larger gold clusters.

In 2004, Chen and Goodman[20] reported kinetic studies of the oxidation of carbon monoxide at gold clusters at a thin titanium dioxide surface grown on to an Mo(112) surface (Figure 9.6). They concluded that the gold bilayer structure is significantly more active (by more than an order of magnitude) than the gold monolayer; the TiO_x is not considered to be directly involved in the bonding of O_2 or CO because the gold overlayer precludes their access to the oxide substrate. It is a contribution of the first and second layers of gold that is necessary to promote the reaction between CO and O_2. In 2003, a group at Ulm took a different approach;[21] they investigated structural, electronic and

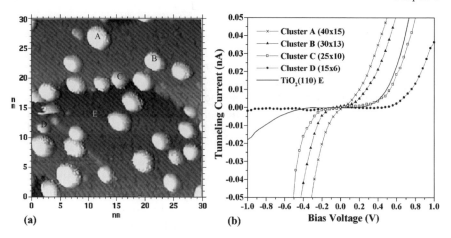

Figure 9.5 STS (current–voltage curves) for gold clusters on TiO_2 (typical STM images shown) of various sizes. (Reproduced from Ref. 5).

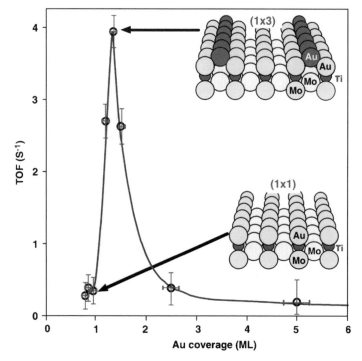

Figure 9.6 Activity for CO oxidation at room temperature as a function of gold coverage on an $Mo(112)$–(8×2)–TiO_x surface. The $CO:O_2$ ratio was 2:1 and the total pressure 5 Torr. Two discrete gold structures were investigated, (1×1) and (1×3). The initial turn over frequencies (TOF) over the (1×1) gold monolayer structure were significantly lower than that for the (1×3) bilayer structure. (Reproduced from Ref. 20).

impurity doping of supported gold nanoclusters by the soft landing of mass-selected Au_n and Au_nSr clusters on to a well-characterised MgO(100) surface. They concluded, as did the Aarhus group, that oxygen vacancies were the nucleation sites for the gold clusters. Also, size-dependent activation of the reactants by the model catalysts resulting in catalytic activity correlated well with electronic structure features of the catalyst, in particular the width and position of the d-band. The doping influence of strontium was also highlighted as providing a significant mechanistic clue (Figure 9.7). That an electronic factor (charge transfer) is significant is very much in accord with the cationic state of gold observed by the Cardiff group and regarded as being crucial to catalytic activity.[22] Their conclusion for the Au–Pd system supported on TiO_2 is that a calcining procedure creates alloy nanoparticles consisting of core gold particles surrounded by palladium; this is somewhat reminiscent of the Ni(111)–Au system found to be active for the dissociation of methane.[12]

There have been a number of reviews which have considered the "new" chemistry associated with gold clusters, but to extract a theme which can be incorporated to develop a unified theory has not been feasible. Meyer et al.[14a] at the Fritz Haber Institute make the point that "experiments seem to be taken in an unsystematic manner so that general trends are therefore often obscured". Hutchings and Haruta[14b] summarised the recent catalytic aspects of gold particles and suggested that it is gold clusters with diameters below 2.0 nm or with less than 300 atoms that will be found to be most fruitful for future developments in heterogeneous catalysis. The recent report[22] of the direct synthesis of hydrogen peroxide from H_2 and O_2 using TiO_2-supported Au–Pd catalysts is a further example of the significance of gold in one form or another in catalysis. On the other hand, gold supported on Fe_2O_3 – under carefully controlled conditions of calcination – is active in the selective oxidation of carbon monoxide in the presence of H_2, CO_2 and water vapour. In the latter system, it was concluded that the most selective catalysts comprise relatively large Au nanocrystals (>5 nm) supported on a reducible oxide.[23]

9.5 Oxidation and Polymerisation at Pd Atoms Deposited on MgO Surfaces

Size-selected palladium atoms were deposited on an *in situ*-prepared MgO(100) thin film at 90 K; the palladium surface concentration was about 1% of a monolayer. Comparison of *ab initio* calculations and FTIR studies of CO adsorption provided evidence for single Pd atoms bond to F centres of the MgO support with two CO molecules attached to each palladium atom.[24]

Preadsorption of oxygen followed by subsequent exposure to CO at 90 K leads to the formation and desorption of CO_2 at 260 and 500 K, suggesting that two reaction mechanisms are involved. What is also significant is that pre-adsorbing CO followed by oxygen leads to no CO_2 formation. The authors suggested the formation of a $Pd(CO)_2O_2$ complex, the transition state involving one of the CO molecules approaching the closest O atom of the oxygen

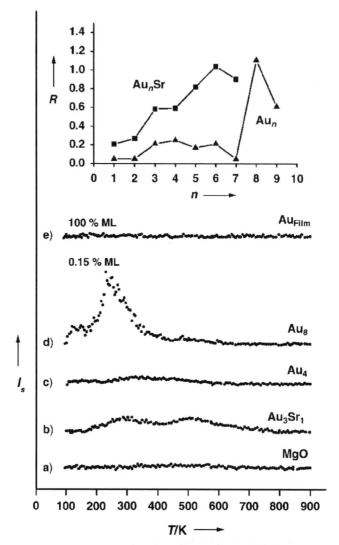

Figure 9.7 Temperature-programmed reaction (TPR) spectra for CO oxidation at a series of model catalysts prepared by the soft landing of mass-selected Au_n and Au_nS_r cluster ions on MgO(100) thin films which are vacancy free (typically 1% of a monolayer). (a) MgO; (b) Au_3Sr; (c) Au_4; (d) Au_8. Also shown is the chemical reactivity R of pure Au_n and Au_nSr clusters with $1 \leq n \leq 9$. (Reproduced from Ref. 21).

molecule, the desorbing CO_2 molecules carrying away the majority of the reaction heat (2.2 eV) partitioned as 0.1, 0.1 and 1.8 eV into the translational, rotational and vibrational degrees of freedom, respectively. In other words, the CO_2 molecule is vibrationally hot. What is also clear is that the O_2 molecule is activated when present at the F centre of the MgO surface. There are some similarities with coadsorption studies of CO and O_2 at aluminium surfaces at

80 K, where carbonate formation occurred but, rather than decompose to desorb as CO_2, the surface carbonate was reduced at the aluminium surface driven by the thermodynamically favoured reaction, forming "oxide" and carbidic carbon.[25] This reaction does not occur at aluminium surfaces at room temperature, presumably because the transition state (complex) did not form, surface oxide formation being the dominant fast reaction pathway. We suggested, however, in contrast to the Ulm group, that the active oxygen state was atomic rather than molecular, although unambiguous evidence was not available. That dioxygen complexes could under some circumstances control reaction pathways, has, however, been established by spectroscopic kinetic studies of the ammonia–dioxygen reaction at Zn(0001) and Ag(111), where in the absence of ammonia the probability of dioxygen cleavage is small but increases by a factor of about 10^3 when coadsorbed with ammonia.[26] The dioxygen–ammonia complex provides an energetically facile route to the formation of surface amide and oxide. It is an example of precursor-assisted bond cleavage.

For CO and NO coadsorption at palladium particles, it was shown that clusters up to Pd_4 were inert for the oxidation reaction leading to CO_2 at 300 K. The authors suggested[27] that the reaction involves oxygen atoms generated by the dissociation of NO. The Cardiff group,[28] using a combination of XPS and HREELS, showed that in the coadsorption of CO and NO at Ag(111) a surface complex, "CO–NO", was formed at low temperature (80 K) but which decomposed at 170 K; the frequencies of the loss features attributed to NO, at $1270 \, cm^{-1}$, were, however, some $400 \, cm^{-1}$ less than those assigned by Heiz and co-workers[27] for NO at Pd_{30}.

The influence of palladium cluster size in the range $1 \leq n < 30$ at MgO(100) thin films was striking.[29] In a single-pass heating cycle experiment, conversion of acetylene to benzene, butadiene and butane was catalysed with different selectivities as a function of cluster size. Up to Pd_3 only benzene is catalysed. The highest selectivity for the formation of butadiene is observed for Pd_6, whereas Pd_{20} is the most selective for butane (Figure 9.8). These results clearly suggest that it might be possible to tune atom by atom the activity and selectivity of real catalysts – but not yet! Although the palladium clusters are well characterised in the gas phase, there is no guarantee that they retain their configurations after being accommodated at the surface. There is nevertheless a striking resemblance to the observation of Ormerod and Lambert,[30] where structure sensitivity for the cycloisomerisation of acetylene was observed with real palladium catalysts.

9.6 Clusters in Nanocatalysis

Thomas and his colleagues in Cambridge have pioneered the development of nanoparticles prepared from cluster compounds and supported in mesoporous silica.[31] Highly active and effective catalysts have been developed for a number of hydrogenation reactions. The significant factors controlling

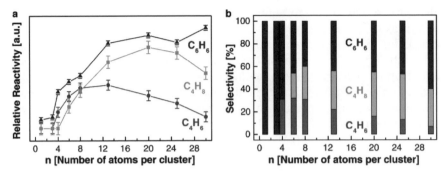

Figure 9.8 Relative reactivities (a) and selectivities (b) of palladium cluster size in the formation of C_6H_6, C_4H_8 and C_4H_6 in the polymerisation of acetylene. (Reproduced from Ref. 29).

activity are their size and the low coordination of the metal atoms involved in the nanoparticles.

The strategy used in catalyst preparation is to start with a mixed metal carbonylate[32] anchored to the surface silanol groups within the inner walls of mesoporous silica and involving $M_mCO\cdots HOSi$ hydrogen bonding or as in the case of the anion $[Ru_6C(CO)_{16}Cl_3]^-$ through an SnCl–HOSi interaction. That the carbonyl group is retained is demonstrated by FTIR spectroscopy, with uniform distribution being achieved with a high surface area (500–900 $m^2 g^{-1}$) silica. Heating these carbonylate clusters in vacuum results in desorption of carbon monoxide and also the organic cationic material, leading to the formation of well-dispersed "naked" bimetallic particles. The authors emphasise that the active site can be identified with FTIR, X-ray absorption and various kinds of scanning transmission electron microscopies. Although the precise way in which the nanoparticles are attached to the internal surface of the mesoporous solid has not been identified with certainty, it is suggested that for the oxophilic metals such as Cu (in Cu_4Ru_{12}) and Ag (in Ag_3Ru_{10}) metal–oxygen bond formation is involved. A remarkable property of these nanoparticles is that they are resistant to sintering under reaction conditions; they retain their positions during a catalytic cycle, they are readily accessible within the mesopores and the tunnel-like nature of the support (MCM-41) ensures that the reactants come into contact with the active sites as they diffuse through the mesopores.

Among the many examples studied by the Cambridge–Royal Institution group is the hydrogenation of 1,5-cyclododecatriene to 1,5-cyclododecadiene, cyclododecene and cyclododecane, which is important in the synthesis of intermediates which are used in the synthesis of nylon-12, polyesters and carboxylic acids. Another example is the conversion of *trans*-muconic acid to adipic acid using nanocatalysts based on Pd_6Ru_6, Cu_4Ru_{12}, Pt_2Ru_{10} and $PtRu_5$ (Figure 9.9).

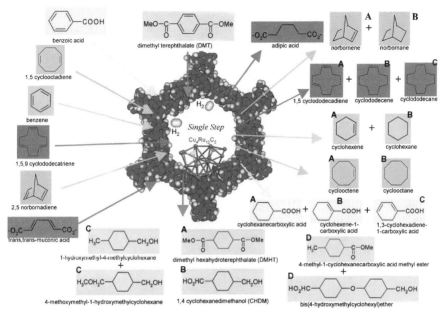

Figure 9.9 Examples of single-step hydrogenations using $Cu_4Ru_{12}C_2$ nanocatalysts. The copper atoms are bonded to the inner wall of MCM-41. (Reproduced from Ref. 32).

9.7 Molybdenum Sulfide Nanoclusters and Catalytic Hydrodesulfurisation Reaction Pathways

A major objective of nanocatalysis is to isolate, at the atom resolved level, the active sites in an individually significant catalytic reaction. Molybdenum-based hydrodesulfurisation (HDS) catalysts are one of the most important to be used in oil refineries. The general view, based on EXAFS studies, is that the active molybdenum is present as small MoS_2 nanostructures and that it is sulfur vacancies at the edges of these structures that are important for the adsorption of the sulfur-containing molecules which need to be removed from the fuels. Haldor Topsøe and the Aarhus group have made very significant advances, both experimentally and theoretically, in understanding the various MoS_2 structures which can form under both reducing and sulfiding reaction conditions.[33] In particular, they have based their conclusions on studies of single-layer MoS_2 nanoclusters which can be formed by sulfiding molybdenum deposited on an Au(111) surface in a sulfiding atmosphere. These nanoclusters of MoS_2 were used as model HDS catalysts for study by STM, using various mixtures of H_2S and H_2 for the synthesis of the sulfide. The exact composition of the gaseous atmosphere used has a significant influence on the cluster morphology. When prepared under the most sulfiding conditions, the STM images indicate the formation of triangular-shaped clusters, suggesting that these are the equilibrium form of MoS_2 clusters under these conditions. Under more reducing conditions ($H_2S:H_2 = 0.07$) there is a dominance of crystals with a hexagonal morphology. However, although the number of edge sites remains relatively unchanged, there is a dramatic change in the type of edges exposed. For example, one of the edges observed in the hexagonal structure (sulfiding conditions) is absent in the triangular structure (reducing conditions). Previous models without the advantage of STM have assumed a hexagonal structure irrespective of the sulfiding conditions, but the evidence for morphology changes at the atom resolved level is likely to influence the catalysis taking place in that large changes in both the absolute and relative concentrations of different types of edge sites (sulfur monomers, dimers, SH groups, vacancies, metallic brim sites, *etc.*) can occur.

The authors compared their STM evidence with DFT calculations and simulations.[34] Whereas the triangular MoS_2 nanoclusters formed under sulfiding conditions were shown to be terminated by fully saturated Mo edges, the hexagonal clusters expose two different edges: Mo edges covered by S monomers and fully saturated S edges with H atoms adsorbed (*i.e.* SH groups). The latter are believed to be important in HDS as a source of H atoms. For every type of MoS_2 edge observed, it is concluded that the electronic structure is dominated by metallic one-dimensional edge states. These were considered to have a significant role in the catalysis and were described as "brim sites".

The authors then proceeded to examine the chemistry associated with these different structures by studying the adsorption of thiophene (C_4H_4S) and were able to pinpoint the precise sites on the MoS_2 nanoclusters where the thiophene adsorbs and reacts. It is found that in the presence of hydrogen, thiophene is hydrogenated and broken down on the fully sulfided MoS_2 clusters, normally regarded as inactive. Sulfur vacancies are not involved. The activity is shown to be associated with the presence of special one-dimensional electronic edge states responsible for the metallic character of the MoS_2 nanoclusters. These so-called "brim sites", in contrast to the inactive and insulating basal plane of MoS_2, have the ability to donate and accept electrons just like catalytically active metals. The reaction leads to the formation of adsorbed thiolate (R–S) intermediates, which are very reactive and readily desulfurised; it is the first step in the hydrodesulfurisation of thiophene. The recognition of these "brim sites" is suggested to be key in the understanding of the hydrogenation of aromatics in general and important in the oil industry. Figure 9.10 shows STM images of a triangular single layer of an MoS_2 nanocluster bonding thiophene at low temperatures. Below 200 K (a), thiophene bonds in two configurations – on top of the "brim sites" associated with the edge states (Type B) and also at the perimeter of the cluster (Type A). Between 200 and 240 K (b), the thiophene molecules present at the "brim sites" have desorbed but those adsorbed at the perimeter edges are still present. Above 240 K, no thiophene molecules are present.

When the nanoclusters are pretreated with atomic hydrogen, a much stronger chemisorbed state of adsorbed thiophene is present. "Beam"-like structures are observed protruding about 0.4 Å above the basal plane in the

Nanoparticles and Chemical Reactivity

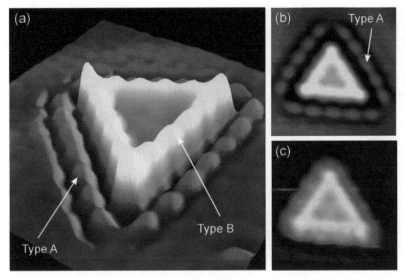

Figure 9.10 STM images of a triangular single-layer MoS$_2$ nanocluster showing the adsorption of thiophene at low temperatures. (a) Below 200 K there are two states, both molecular, one adsorbed on top of the bright rim associated with an edge (Type B) and the other adsorbed at the perimeter of the nanocrystal (Type A); in (b), only Type A exists between 200 and 240 K; (c) above 240 K no thiophene is present. (Reproduced from Ref. 33).

row adjacent to the bright brim of the MoS$_2$ cluster. This is also made clear where two line scans obtained from the edges before and after thiophene reaction with hydrogen are compared (Figure 9.11). The authors develop detailed arguments and conclude that the adsorbed species are coordinated to the fully sulfided Mo edges and that the molecules must be intermediates formed in a reaction at the "brim sites". Since these are only observed on the H-treated clusters, the authors propose that SH groups are formed and that these play an essential role in the reaction with thiophene at sites present on the metallic brim.

9.8 Nanoparticle Geometry at Oxide-supported Metal Catalysts

In 2001, an experimental approach for investigating the structures of nanoparticles was described by the group in the Department of Materials Science and Metallurgy in Cambridge.[35,36] It was based on a variant of three-dimensional electron microscopy, Z-contrast tomography, where using a high-resolution transmission electron microscope (HRSTEM) equipped with a high-angle annular dark-field detector (HAADF) images are formed by

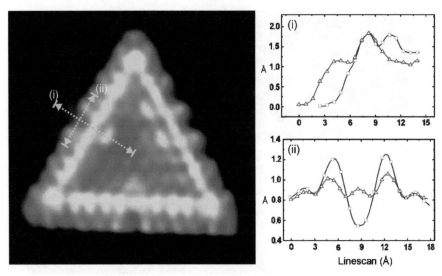

Figure 9.11 Thiophene adsorbed at 500 K on an H-atom pretreated MoS_2 cluster (50×54 Å^2). Beam-like features at the metallic edge [scan line (i)] and the shifted intensity of the outermost edge protrusions relative to the clean edge (triangles refer to the clean edge). These shifts in intensity [line scan (ii)] are associated with changes in the local electronic structure after adsorption of thiophene observed with STM. All the images were taken at room temperature subsequent to thiophene adsorption at 500 K. (Reproduced from Ref. 34).

Rutherford scattered electrons (Figure 9.12). The intensity of these electrons is proportional to Z^2 (where Z is the atomic number of scattering atom) so that the experimental method is most suitable for high-Z materials distributed over low-Z supports.

Earlier, Thomas and his group at the Royal Institution had in 1986, using HRSTEM, obtained information on the two-dimensional (2D) local picture of minute particles of metals supported on high-area oxides and carbon. But the breakthrough of being able to obtain 3D images providing information on the topography of nanoparticles gave a new insight to the factors that control activity and selectivity in heterogeneous catalysis (Figure 9.13). The authors predicted that "the techniques should be of enormous benefit to the catalyst community". Thomas' long-time interest in bimetallic nanocatalysts, active in selective hydrogenation reactions under solvent-free conditions, was an obvious area for exploring how their approach could elucidate such information as their composition, shape, location and distribution within the mesoporous host. In particular, they established that for $Ru_{10}Pt_2$ supported on mesoporous silica under "bright field" conditions the particles are barely noticed, but under HAADF conditions the particles are seen clearly.[37] They established that for particles of 1 nm or less in size HAADF tomography is essential. In 2004, Thomas and Midgley[38] described, in an authoritative review, the important developments on how electron-optical methods have provided unique insights

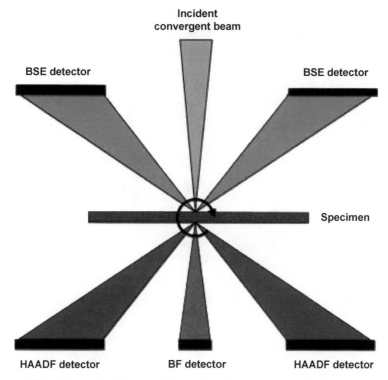

Figure 9.12 Schematic diagram illustrating the geometry of detectors used for STEM BF, STEM HAADF and STM BSE imaging. (Reproduced from Ref. 35).

in solid-state and materials chemistry, with emphasis given to their relevance to surface reactivity and catalysis.

Although model systems have provided new insights into the nature of the active site in heterogeneously catalysed reactions, supported catalysts, as used in industry, are not as well defined as perhaps envisaged in model systems. Janssens and co-workers[39] at Haldor Topsøe have recently described atomic-scale geometric models for supported fcc metal nanoparticles from the measurement of particle sizes and particle volume by scanning transmission electron microscopy (STEM) and metal–metal coordination numbers determined from EXAFS. They chose gold particles supported on TiO_2, $MgAl_2O_3$ and Al_2O_3. The models enable estimates to be made of geometric properties such as specific gold surface area, metal–support contact perimeter and area, edge length and the number of gold atoms located at the corners of the particles. The gold content of the supported catalysts were all approximately 4 wt%; they were all treated in a similar fashion – heated for 1 h in a mixture of 1% CO, 21% O_2 and 78% Ar at atmospheric pressure. Subsequently they were exposed to the gas mixture at 0 °C. The gold particle sizes and volumes for the three catalysts were derived from a high-angle annular dark field (HAADF) detector attached to an

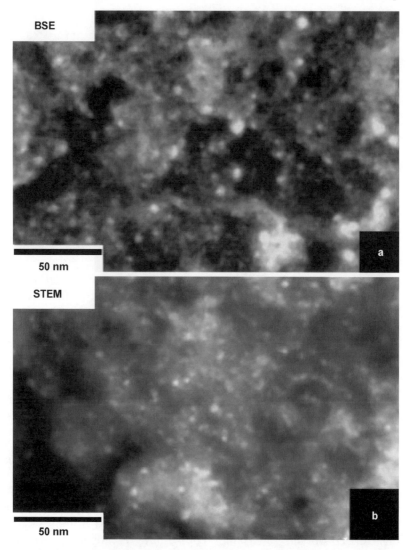

Figure 9.13 (a) SEM BSE image and (b) STEM HAADF image of Pd nanoparticles on a carbon support. The clarity of the images illustrates the advantage of the HAADF and BSE approach. (Reproduced from Ref. 36).

STEM. The particle size distributions are shown in Figure 9.14; they are very different for the three catalysts. The average gold particle sizes are 2.1, 3.5 and 1.6 nm for the TiO_2-, $MgAl_2O_4$- and Al_2O_3- supported catalysts, respectively. Clearly, those at Al_2O_3 are the smallest; those at TiO_2 are smaller than those supported at $MgAl_2O_4$ and also thicker. Differences in metal–support interfacial energies are suggested to account for the variation in particle size and shape.

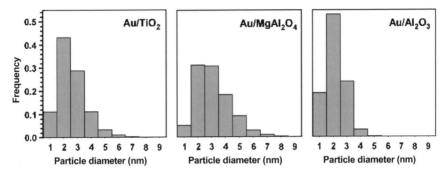

Figure 9.14 Particle size distribution of the Au/TiO_2, $Au/MgAl_2O_4$ and Au/Al_2O_3 catalysts obtained from STEM images. (Reproduced from Ref. 39).

The authors then went further and related the size and shape of the catalysts to their activities in the oxidation of carbon monoxide.[40] A single geometric factor was clearly not involved in controlling activity. Taking evidence from DFT calculations and IR activity for the significance of low-coordinated edge atoms in CO oxidation over gold, the authors proceeded to estimate the turnover frequency per corner gold atom and for the Au/TiO_2 and $Au/MgAl_2O_4$ catalysts they find the same value of $0.8\,s^{-1}$ even though the gold particle shapes are different. This means that the different overall activities of Au/TiO_2 and $Au/MgAl_2O_4$ can be accounted for by the requirement for low-coordinated gold atoms. The turnover frequency for the Au/Al_2O_3 catalyst was about four times smaller so that other effects are involved and probably related to the support. Janssens and colleagues also argue that the suggestion that activity is related to a two-layer structure, as proposed by Chen and Goodman,[41] cannot account for their observations, as only about 1% of the gold is present in particles with two layers in the Au/TiO_2 catalyst.

9.9 Summary

The preparation, morphology and stability of nanoparticles have dominated much of the STM studies over the last decade, with gold particles supported on TiO_2 attracting particular attention. Relating their chemical reactivity with any particular common factor has not, however, emerged. Information concerning the 3D nature of nanoparticles has also been obtained from experimental advances in STEM. The encapsulation of nanoparticles within porous oxides has been an area where Thomas and his colleagues[38] have taken advantage of high-resolution transmission electron microscopy – which they describe as the "ultimate nanoanalytical technique" – to determine the elemental composition and morphology of particles consisting of no more than a dozen atoms, such information being essential for the development of their views on single-site catalysis.

Studies of gold particles supported on oxides have been widespread[14] in an attempt to find the origin of gold's unusual and unexpected catalytic oxidation activity, bilayer structures,[20] edge sites[40] and cationic states[14b] being proposed by various research groups but with one feature, namely that catalytic activity is usually (but not always[23]) confined to particles less than about 2 nm in size. Only in a relatively few cases has more detailed information become available. A recent paper (2006) by Overbury et al.[42] concludes that the activity of gold catalysts is "sensitive to many factors that may mask the true structure dependence" and that "the observed decrease in activity with increasing particle size beyond 2 nm is controlled by the population of low-coordinated sites rather than by size-dependent changes in overall electronic structure of the nanoparticle".

In the case of palladium particles supported on magnesium oxide, Heiz and his colleagues have shown,[29] in an elegant study, a correlation between the number of palladium atoms in a cluster and the selectivity for the conversion of acetylene to benzene, butadiene and butane, whereas in the industrially significant area of catalytic hydrodesulfurisation, the Aarhus group,[33] with support from theory, have pinpointed by STM metallic edge states as the active sites in the MoS_2 catalysts.

As to the number of atoms required to "close the gap" between insulator and metallic clusters, they vary from as few as 20 to several hundred atoms. Freund[3] suggests that the precise numbers will vary from metal to metal, depending on the electronic structure of the metal.

References

1. M. W. Roberts, A. F. Carley and S. R. Grubb, *J. Chem. Soc., Chem. Commun.*, 1984, 459.
2. A. F. Carley, P. R. Chalker and M. W. Roberts, *Proc. R. Soc. London, Ser, A*, 1985, **399**, 167.
3. H. -J. Freund, *Surf. Sci.*, 2002, **500**, 271.
4. C. Xu, X. Lai and D. W. Goodman, *Faraday Discuss.*, 1996, **105**, 247.
5. D. C. Meier, X. Lai and D. W. Goodman, in *Surface Chemistry and Catalysis*, ed. A. F. Carley, P. R. Davies, G. J. Hutchings and M. S. Spencer, Kluwer Academic/Plenum Press, New York, 2002, 148.
6. S. J. Caroll, S. Pratonep, M. Streun, R. E. Palmer, S. Hobday and R. Smith, *J. Chem. Phys.*, 2000, **113**, 7723.
7. S. Pratonep, S. J. Carroll, C. Xirouchaki, M. Streun and R. E. Palmer, *Rev. Sci. Instrum.*, 2005, **76**, 045103.
8. J. W. A. Sachtler, J. P. Biberian and G. A. Sormorjai, *Surf. Sci.*, 1981, **110**, 43.
9. V. Ponec and G. C. Bond, *Catalysis by Metals and Alloys*, Elsevier, Amsterdam, 1995, Chapter 10.
10. W. A. A. van Barneveld and V. Ponec, *J. Catal.*, 1978, **51**, 426.

11. K. Kishi and M. W. Roberts, *J. Chem. Soc., Faraday Trans. 1*, 1975, **71**, 1715; R. Joyner and M. W. Roberts, *J. Chem. Soc., Faraday Trans. 1*, 1974, **70**, 1819.
12. F. Besenbacher, I. Chorkendorff, B. S. Claussen, B. Hammer, A. M. Molenbroek, J. K. Norskov and I. Stensgaard, *Science*, 1998, **279**, 1913.
13. M. B. Hugenschmidt, A. Hitzake and R. J. Behm, *Phys. Rev. Lett.*, 1996, **76**, 2535.
14. See the following reviews: (a) R. Meyer, C. Lemire, Sh. K. Shaikhutdinov and H. -J. Freund, *Gold Bull.*, 2004, **37**, 72; (b) G. J. Hutchings and M. Haruta, *Appl. Catal. A*, 2005, **291**, 2.
15. E. Wahlström, N. Lopez, R. Schaub, P. Thostrup, A. Rønnau, C. Africh, E. Laesgaard, J. K. Norskov and F. Besenbacher, *Phys. Rev. Lett.*, 2003, **90**, No. 2, 026101-1.
16. C. E. J. Mitchell, A. Howard, M. Carney and R. G. Edgell, *Surf. Sci.*, 2001, **490**, 196.
17. R. Meyer, M. Bäumer, Sh. K. Shaikhutdinov and H.-J. Fruend, *Surf. Sci.*, 2003, **546**, L813.
18. R. Meyer, D. Lahav, T. Schalow, M. Laurin, B. Brandt, S. Schauermann, S. Guimond, T. Klüner, H. Kuhlenbeck, J. Libuda, Sh. K. Shaikhutdinov and H. -J. Freund, *Surf. Sci.*, 2005, **586**, 174.
19. D. E. Starr, Sh. K. Shaikhutdinov and H.-J. Freund, *Top. Catal.*, 2005, **36**, 33.
20. M. S. Chen and D. W. Goodman, *Science*, 2004, **306**, 252.
21. H. Häkkinen, S. Abbet, A. Sanchez, U. Heiz and U. Landman, *Angew. Chem. Int. Ed.*, 2003, **42**, 1297.
22. J. K. Edwards, B. E. Solsona, P. Landon, A. F. Carley, A. Herzing, C. J. Kiely and G. J. Hutchings, *J. Catal.*, 2005, **236**, 69.
23. P. Landon, J. Ferguson, B. E. Solsona, T. Garcia, A. F. Carley, A. A. Herzing, C. J. Kiely, S. E. Golunski and G. J. Hutchings, *Chem. Commun.*, 2005, 3385.
24. S. Abbert, U. Heiz, H. Häkkinen and U. Landman, *Phys. Rev. Lett.*, 2001, **86**, 5950.
25. A. F. Carley and M. W. Roberts, *J. Chem. Soc., Chem. Commun.*, 1987, 355.
26. A. F. Carley, S. Yan and M. W. Roberts, *J. Chem. Soc., Chem. Commun.*, 1988, 267; *J. Chem. Soc., Faraday Trans.*, 1990, **86**, 2701.
27. A. S. Wörz, K. Judai, S. Abbet and U. Heiz, *J. Am. Chem. Soc.*, 2003, **125**, 7964.
28. A. F. Carley, P. R. Davies, M. W. Roberts, A. K. Santra and K. K. Thomas, *Surf. Sci.*, 1998, **406**, L587.
29. S. Abbet, A. Sanchez, U. Heiz and W. -D. Schneider, *J. Catal.*, 2001, **198**, 122.
30. R. M. Ormerod and R. M. Lambert, *J. Chem. Soc., Chem. Commun.*, 1990, 1421.
31. J. M. Thomas, *Faraday Discuss.*, 1996, **105**, 1.
32. B. F. G. Johnson, *Top. Catal.*, 2003, **24**, 147.

33. J. V. Lauritsen, M. Nyberg, J. K. Norskov, B. S. Clausen, H. Topsøe, E. Laegsgaard and F. Besenbacher, *J. Catal.*, 2004, **224**, 94.
34. M. V. Bollinger, J. V. Lauritsen, K. W. Jacobsen, J. K. Norskov, S. Helveg and F. Besenbacher, *Phys. Rev. Lett.*, 2001, **87**, 196803.
35. P. A. Midgley, M. Weyland, J. M. Thomas and B. F. G. Johnson, *Chem. Commun.*, 2001, 907.
36. M. Weyland, P. A. Midgley and J. M. Thomas, *J. Phys. Chem. B*, 2001, **105**, 7882; P. A. Midgley, M. Weyland, J. M. Thomas, P. L. Gai and E. D. Boyes, *Angew. Chem. Int. Ed.*, 2002, **41**, 3804.
37. J. M. Thomas, P. A. Midgley, T. J. V. Yates, J. S. Barnard, R. Raja, I. Arslan and M. Weyland, *Angew. Chem. Int. Ed.*, 2004, **43**, 6745.
38. J. M. Thomas and P. A. Midgley, *Chem. Commun.*, 2004, 1253.
39. A. Carlsson, A. Puig-Molina, T. V. W. Janssens, *J. Phys. Chem. B*, 2006, **110**, 5286.
40. T. W. Janssens, A. Carlsson, A. Puig-Molina and B. S. Clausen, *J. Catal.*, 2006, **240**, 108.
41. M. S. Chen and D. W. Goodman, *Science*, 2004, **306**, 252.
42. S.H. Overbury, V. Schwartz, D.R. Mullins, W. Yan and S. Dai, *J. Catal.*, 2006, **241**, 56.
43. W. D. Luedke and U. Landman, *Phys. Rev. Lett.*, 1999, **82**, 3835.

Further Reading

J. Grunes, J. Zhu and G. A. Somorjai, Catalysis and nanoscience, *Chem. Commun.*, 2003, 2257.

J. M. Thomas, C. R. Catlow and G. Sankar, Determining the structure of active sites, transition states and intermediates in heterogeneously catalysed reactions, *Chem. Commun.*, 2002, 2921.

P. M. Holblad, J. H. Larsen, I. Chorkendorff, L. P. Nielsen, F. Besenbacher, I. Stensgaard, E. Laesgaard, P. Kratzer, B. Hammer and J. K. Norskov, Designing surface alloys with specific active sites, *Catal. Lett.*, 1996, **40**, 131.

J. B. Park, J. S. Ratliff, S. Ma and D. A. Chen, *In situ* STM studies of bimetallic cluster growth: Pt–Rh on TiO_2 (110), *Surf. Sci.*, 2006, **600**, 2913.

M. Haruta, Catalysis of gold nanoparticles desposited on metal oxides, *Cattech*, 2002, **6**, 102.

K. Tanaka and Z -X. Xie, Composite nano-structures controlled by weak interactions on solid surfaces, *Catal. Lett.*, 2002, **19**, 149.

H. -J. Freund, Metal-supported ultrathin oxide film systems as designable catalysts and catalyst supports, *Surf. Sci.*, 2007, **601**, 1438.

M. W. Roberts, The nature and reactivity of chemisorbed oxygen and oxide overlayers at metal surfaces as revealed by photoelectron spectroscopy, in *Structure and Reactivity of Surfaces*, ed. C. Morterra, A. Zechina and G. Costa, Elsevier, Amsterdam, 1989, p. 787.

CHAPTER 10
Studies of Sulfur and Thiols at Metal Surfaces

"Nothing has such power to broaden the mind as the ability to investigate systematically and truly all that comes under your observation in life"

Marcus Aurelius

10.1 Introduction

Sulfur is a natural contaminant of fossil fuels and poses a severe problem to today's technology in two main respects: the formation of sulfur oxides when fuels are burnt and the poisoning of catalysts. The former contributes to the acidification of rain water, causing corrosion and killing vegetation, and the latter results from the high affinity of sulfur for nearly all metals, even gold, and is a problem in reforming, Fischer–Tropsch and exhaust catalysts. The dramatic effect of sulfur poisoning on iron Fischer–Tropsch catalysts is shown in Figure 10.1.[1]

Environmental legislation has imposed increasingly stringent limits on the sulfur content of fuels; in Europe, for example, permitted sulfur levels have been reduced from 3000 ppm in 1990 to less than 10 ppm by 2008, and similar legislation is in place in the USA, Japan and Australia. Much of the research into sulfur and sulfur-containing molecules at surfaces therefore follows two main themes: understanding the mechanism of catalyst poisoning and studying hydrodesulfurisation catalysis (HDS) to improve sulfur removal methods. However, although the emphasis is mainly on prevention, the presence of sulfur at a surface can also have a positive effect, either as a promoter[2] or as a selective poison to eliminate an undesirable pathway.

Another important reason for studying sulfur and sulfur-containing molecules at surfaces is the increasing interest in self-assembled films. The affinity of the thiol group for metals and in particular the otherwise inert gold surface has led to a large number of studies into the structure of these systems. The complex local structures frequently seen in these systems are difficult to interpret through diffraction methods, but have proved to be an ideal testing ground

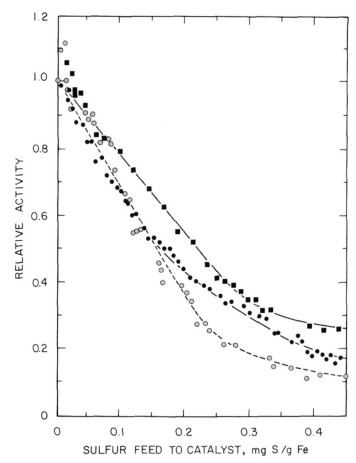

Figure 10.1 Poisoning of a fused iron catalyst by H_2S at 535 K. $H_2/CO = 1$, $P = 2.16$ MPa. Sulfur concentration in feed $(mg\,S\,m^{-3})$ (○) 6.9, (●) 23.0, (■) 69.0. (Reproduced from Ref. 1).

for scanning probe microscopies. In this chapter, we consider the changes in surface structure caused by sulfur adlayers and discuss the contribution made to our understanding of the systems by STM. We also briefly summarise some of the major results reported in the study of thiol groups related to self-assembled films, but we do not deal with this area comprehensively because of the extensive literature available.

10.2 Studies of Atomic Sulfur Adsorbed at Metal Surfaces

The structural behaviour of sulfur at metal surfaces has proven to be very rich. Large-scale reconstruction has been proposed in many cases and models for several of these systems still remain contentious. An early pioneer of the area was J. Oudar in Paris, who, in the 1970s, used LEED and radioactive sulfur studies (as H_2S and S vapour) to describe the major structural phases present at a number of metal surfaces as a function of coverage. More recently, the availability of techniques such as surface X-ray diffraction (SXRD) and surface extended X-ray absorption fine structure (SEXAFS). which can provide detailed information on the inside of the unit cell, has helped clarify many of the issues raised by the early models. STM has also played a role here, although the information that it provides about the unit cell has perhaps been less important than *in situ* studies of mass transfer during reaction. The latter has contributed significantly to our understanding of the mechanisms of the structural changes that occur when sulfur reacts with a metal surface.

In the sections that follow, we describe the evolution of models for the major phases of sulfur at a number of surfaces, beginning with copper and nickel, which show many similarities.

10.2.1 Copper

All three of the copper basal plane surfaces show complex structural changes as the surface concentration of sulfur is changed. With the Cu(111) surface, Domange and Oudar[3] reported a $(\sqrt{3} \times \sqrt{3})\ R30°$ structure after exposure to H_2S at room temperature. On further exposure, this phase was replaced by a $(7\sqrt{} \times \sqrt{7})\ R19°$ structure via a "complex phase". The latter has been identified by more recent studies as a $\begin{pmatrix} 4 & 1 \\ -1 & 4 \end{pmatrix}$ surface mesh and the final phase confirmed by both STM and LEED. However, the $(\sqrt{3} \times \sqrt{3})\ R30°$ structure has not been observed subsequently. Possibly it exists in a very narrow range of surface concentrations. Detailed structural models have been derived for all the phases, principally from STM[4] and SXRD[5] data, and show that in both cases the sulfur causes a surface reconstruction to form Cu_4S tetramers.[6] A model of the $(\sqrt{7} \times \sqrt{7})\ R19°$ structure and STM images of coexisting domains are shown in Figure 10.2.[7]

The initial adsorption of sulfur on the Cu(100) surface leads to a $p(2 \times 2)$ structure[3] for coverages up to 0.25 monolayers, with the sulfur adatom present in the fourfold hollow sites and causing a small outward substrate relaxation.[8–10] A $c(2 \times 2)S$ structure has been reported[11] after adsorption of H_2S at low temperature ($<125\,K$), but is not seen for adsorption at room temperature; higher sulfur concentrations under these conditions lead to a more diffuse LEED pattern.[12] Colaianni and Chorkendorff's STM studies[12] of Cu(100) surfaces at room temperature show a roughening of the terraces during H_2S adsorption but no discernible step-edge movement. Their interpretation is that the sulfur is extracting copper atoms from the terraces and forming a new overlayer, which develops from small island structures. Annealing of this high coverage of sulfur results in the formation of the well-ordered $(\sqrt{17} \times \sqrt{17})\ R14°$ adlayer identified by Domange and Oudar (Figure 10.3[12]). The STM images show a well-ordered structure consisting of units of four atoms.

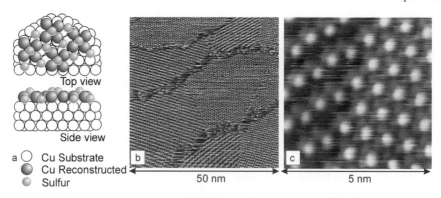

Figure 10.2 Adsorbed sulfur structures on Cu(111). (a) Model of the $(\sqrt{7} \times \sqrt{7})$ R19° phase showing the Cu$_4$S tetramers; large grey circles are added coppers, smaller circles represent S. (b) Filtered 50 × 50 nm STM image of coexisting $(\sqrt{7} \times \sqrt{7})$ R19° and "complex" structures. (c) 5 × 5 nm STM image of domain boundary between the two phases. (Reproduced from Refs. 6 and 7).

In the light of these data, a revision of Domange and Oudar's original model[3] has been suggested[12] (Figure 10.3c–e). Colaianni et al.[13] have used oxygen preadsorption to generate higher sulfur coverages via chemisorptive replacement with desorption of water. A new $\begin{pmatrix} 5 & 2 \\ 0 & 5 \end{pmatrix}$ structure is observed under these conditions characterised by a zig-zag appearance in the STM images that arises from domains of the mirror image structure.

Oudar and co-workers studied the dissociative chemisorption of hydrogen sulfide at Cu(110) surfaces between 1968 and 1971.[3,14] As in the case of Ni(110) described below, a series of structures were identified, which in order of increasing sulfur coverage were described as c(2 × 2), p(5 × 2) and p(3 × 2). In contrast to nickel, the formation of the latter phase is kinetically very slow from the decomposition of H$_2$S and could only be produced at high temperatures and pressures. The c(2 × 2) and p(5 × 2) structures were confirmed by LEED,[15–17] but the p(3 × 2) phase has not been observed by H$_2$S adsorption since Oudar and colleagues' work.

A number of structural models were advanced to explain the observed LEED patterns and later STM images of sulfur at a Cu(110) surface. All of these models suggested sulfur adsorption in a combination of both hollow and bridge sites, but this conflicted with the parameters derived from a SEXAFS study of the c(2 × 2) and p(5 × 2) structures,[18] which established a single adsorption site, the "two-fold" hollow, for the sulfur adatom over the entire coverage range encompassed by the major sulfur phases. The conflict was resolved following a study[19] combining XPS and STM which proposed buckled structures for highest coverage structures and defined concentration boundaries for all three phases on the Cu(110) surface: c(2 × 2) ($\sigma_s < 4.4 \times 10^{14}\,\text{cm}^{-2}$), p(5 × 2) ($4.4 \times 10^{14}\,\text{cm}^{-2} < \sigma_s < 6.6 \times 10^{14}\,\text{cm}^{-2}$) and p(3 × 2) ($\sigma_s = 7.1 \times 10^{14}\,\text{cm}^{-2}$). The last was obtained under relatively mild conditions from the thermal decomposition of a chemisorbed thiol adlayer. The latter was obtained from

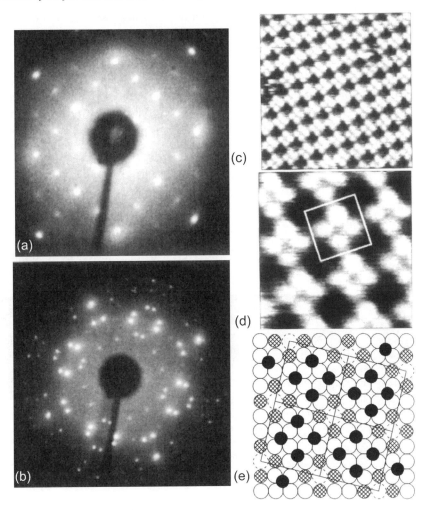

Figure 10.3 Adsorbed sulfur structures on Cu(100). (a, b) LEED patterns from the p(2 × 2) and ($\sqrt{17} \times \sqrt{17}$) R14° structures, respectively. (c) STM image (9.3 × 9.3 nm) of the ($\sqrt{17} \times \sqrt{17}$) R14° structure formed after annealing the sulfur adlayer to 1173 K. (d) High-resolution STM image (2.9 × 2.9 nm) of (c). (e) Proposed model of the ($\sqrt{17} \times \sqrt{17}$) R14° structure; black circles are sulfur adatoms in four-fold sites in the top layer; shaded circles are sulfur adatoms which have replaced a terrace copper atom; dashed circles indicate a copper atom which may be missing. (Adapted from Ref. 12).

the chemisorbed replacement of a preadsorbed oxygen adlayer and decomposed at ~100 °C to give a pure, high-concentration sulfur adlayer.

$$CH_3SH(g) + O(a) \rightarrow 2CH_3S(a) + H_2O(g)$$

$$CH_3S(a) \rightarrow C_2H_6(g) + CH_4(g) + C_2H_4(g) + S(a)$$

From the STM images, the c(2 × 2) structure was shown to consist of ordered domains on average two unit cells wide and separated from each other by 0.36 nm (*i.e.* one substrate unit cell). The expanded section of Figure 10.4b shows the anti-phase nature of these domains, which help to reduce further the surface stress due to the expansion in the lattice caused by sulfur adsorption. The domain boundaries appear to retain some degree of mobility and atomic resolution is very difficult in these areas at room temperature. This was attributed to diffusion of sulfur or copper atoms along the boundaries. Interestingly, domain boundaries can be discerned immediately that islands of the c(2 × 2) structure appear in the STM images (Figure 10.4a). The explanation for this is probably that although the islands of c(2 × 2) look isolated at the surface, they are in fact surrounded by a relatively high concentration of mobile sulfur adatoms that are not imaged by STM. This model is supported by the image of an apparently clean Cu(110) surface obtained by STM when the XPS data show the presence of at least one-third of a monolayer of sulfur. It is also supported by adsorbing oxygen on the adlayer, which has the effect of trapping the sulfur so that it becomes visible in the STM images.[19]

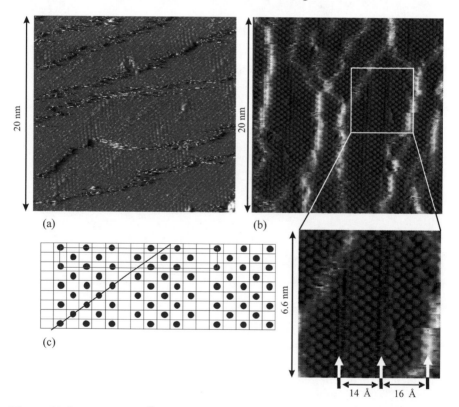

Figure 10.4 STM images[19] of the c(2 × 2)S phase on Cu(110). (a) Islands of c(2 × 2) already showing the discontinuities that characterise this structure. (b) Complete c(2 × 2) phase at a sulfur concentration of 4.4×10^{14} cm^{-2}. (c) Model structure for the c(2 × 2) phase showing domain boundaries.

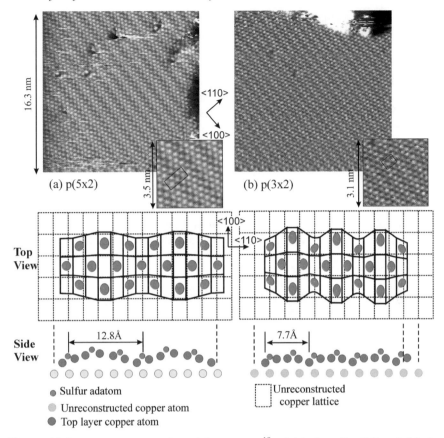

Figure 10.5 STM images and model structures[19] of (a) the p(5 × 2)S and (b) the p(3 × 2)S phases on Cu(110).

Buckled surface models for the two other copper–sulfur phases were proposed that rationalised the STM, LEED and SEXAFS data (Figure 10.5). Note that the local adsorption site of the sulfur adatoms in both the p(3 × 2) and p(5 × 2) structures is the four-fold hollow in agreement with the SEXAFS[18] measurements but looking at the structure from above with the STM, the periodicity of the structure suggests a variety of different adsorption sites.

10.2.2 Nickel

Perdereau and Oudar's early paper[20] reported LEED patterns and surface concentration data for the (111), (100) and (110) surface planes. For Ni(111) a sequence of structures was observed starting with p(2 × 2), changing to $(\sqrt{3} \times \sqrt{3})$ $R30°$ with increasing sulfur concentration and finally to a structure labelled "SBAII", which Edmonds et al.[21] identify as a $(5\sqrt{3} \times 2)$ structure. The low-concentration structures agree well with a model involving sulfur coordination into the highest coordination sites available, the three fold hollow, but the highest concentration has a more complex structure. Perdereau and Oudar's model for the latter was a surface sulfide incorporating nickel into the upper adlayer with the sulfur atoms existing in several different adsorption sites with respect to the underlying (111) lattice. Edmunds et al.,[21] on the other hand, suggested a pseudo-Ni(100)–c(2 × 2)S reconstruction of the surface. From the SEXAFS data of Warburton et al.,[22] it was possible to rule out Perdereau and Oudar's model but, although the match to the pseudo-(100) model was good, the authors expressed some reservations about the quality of the fit. STM studies of the Ni(111)–S system from the group at Aarhus[23] confirmed the c(2 × 2) structure and shed some new light on the situation at higher coverages. The STM images (Figure 10.6), show a well defined $(5\sqrt{3} \times 2)$ phase but, in addition, observations of the development of the structure reveal areas of increased height. These were attributed to islands of nickel formed from atoms ejected from the surface of the terraces due to a 20% decrease in nickel density

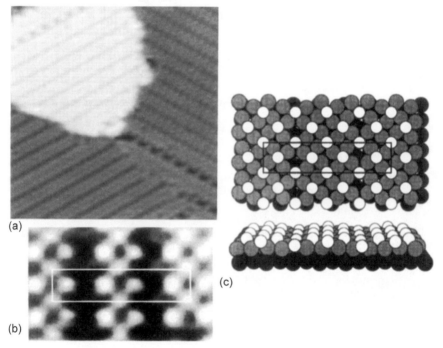

Figure 10.6 STM images of the Ni(111) $(5\sqrt{3} \times 2)$S phase and a model for the structure proposed to explain the decreased density of nickel within the islands. (a) 15.0 × 16.5 nm image showing the three possible domains of the $(5\sqrt{3} \times 2)$S structure; the brighter part of the image corresponds to an adlayer that has developed on top of a nickel island formed during H$_2$S adsorption. (b) 1.8 × 2.9 nm atomically resolved image of the $(5\sqrt{3} \times 2)$S structure. (c) Proposed "clock" structure for the $(5\sqrt{3} \times 2)$S phase that accounts for the reduced nickel density in the sulfur adlayer. (Reproduced from Refs. 23 and 25).

in the sulfur adlayer (step-edges were shown to remain stationary during sulfur adsorption). On the basis of these measurements, Ruan et al.[23] proposed a missing nickel row model for the $(5\sqrt{3} \times 2)$ structure. However, Woodruff pointed out[24] that the Aarhus model disagreed with SEXAFS data, which strongly suggest a four-fold coordination site. The issue has since been resolved by the Aarhus group[25] using grazing incidence X-ray diffraction, from which a "clock" structure was deduced. This retains the four-fold coordination of the sulfur atoms suggested by SEXAFS and the significant height corrugations shown by STM.

In contrast to the (111) plane, the structural details of sulfur adsorbed at Ni(100) surfaces is relatively straightforward, giving a disordered adlayer at low concentrations with patches of $p(2 \times 2)$ developing as the coverage increases evolving to a $c(2 \times 2)$S structure with higher exposures. LEED I/V measurements showed the sulfur adatoms to be adsorbed in the four-fold hollow sites. Partridge et al.[26] studied the influence of sulfur on the oxidation of Ni(100) surfaces and reported that sulfur concentrations as low as 16% of a monolayer inhibited oxygen island nucleation on terraces with islands growing instead at step-edges. The authors do not explain this phenomenon but, in the light of the STM studies of oxidation discussed in Chapter 5, it seems likely that the effect of the sulfur is to prevent oxygen diffusing away from the dissociation sites at the step-edges. A second interesting observation from this study was that oxygen adsorption led to a change in LEED pattern from $p(2 \times 2)$ to $c(2 \times 2)$. The authors discuss the possibility that oxygen is adsorbing into the vacant four-fold sites within the sulfur $p(2 \times 2)$ lattice, but they were unable to verify this observation with STM. An alternative, prompted by more recent observations[19] in Cardiff of the mobility of sulfur adlayers at metal surfaces, is that oxygen adsorption compresses the sulfur structure, resulting in a change from the low- to the high-concentration phase. Similarly, the sulfur adlayer might be expected to compress the oxygen adlayer and accelerate the formation of oxide nuclei, and this was indeed reported by Partridge et al.[26]

Madix and co-workers have also considered the possibility of adsorption at the vacant hollow sites in the $p(2 \times 2)$S adlayer on Ni(110) as part of a study[27,28] of the influence of sulfur on the reaction pathway for alcohol decomposition at nickel surfaces; TPD studies showed[27] complete dehydrogenation to carbon monoxide and hydrogen at clean nickel surfaces, but only partial dehydrogenation to the related aldehydes when sulfur was present. A more recent study[28] using STM considered structural aspects of the system. For sulfur coverages below ~ 0.35 ML, exposure to the alcohols resulted in the adsorption of alkoxides at *defect* sites within the $p(2 \times 2)$S adlayer (Figure 10.7). The change in decomposition pathway of these species to the corresponding aldehydes being brought about by the lack of nearby adsorption sites for the products. No adsorption was observed in the vacant hollow sites. At higher sulfur coverages, where the $c(2 \times 2)$S structure dominates, few vacancies exist within the sulfur adlayer and no alcohol adsorption was observed.

Perdereau and Oudar's study[20] of sulfur adsorption at Ni(110) surfaces showed a very similar behaviour to that of Cu(110). They identified a series of

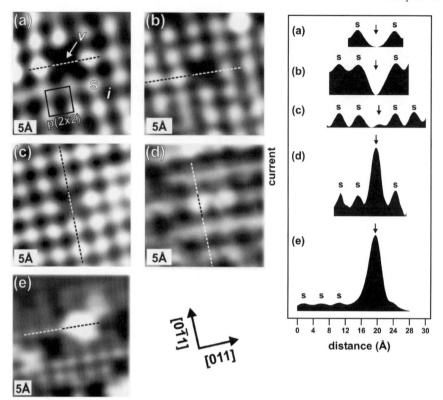

Figure 10.7 Constant-height STM images and line profiles of a partially sulfided Ni(100) surface before (a) and after (b–e) exposure to different alcohols: (a) 0.23 ML sulfur; (b) CH_3OH; (c) CH_3CH_2OH; (d) $CH_3CH_2CH_2OH$; (e) C_6H_5OH. (Reproduced from Ref. 28).

structures at room temperature beginning with $c(2 \times 2)$S, which progressed through $p(5 \times 2)$ and $c(8 \times 2)$ structures to $p(3 \times 2)$. Unlike copper, however, with higher temperature and pressure they were able to obtain a final $p(4 \times 1)$ phase with a sulfur concentration, determined by radioactive sulfur tracer methods, of 0.75 ML. The $c(2 \times 2)$ phase has received most attention in the literature; the sulfur adsorption site in this structure was identified[29] by SEXAFS as the rectangular hollow site, bonding to one Ni atom in the second layer with a 12% expansion of the top layer of the surface from the bulk lattice. The intermediate phases [$p(5 \times 2)$ to $p(3 \times 2)$] have been accepted as simple overlayer structures formed from an increasing compression of the sulfur adlayer in the <110> direction without any reconstruction of the nickel atoms. In the light of the models advanced[19] for the Cu(110) surface, however, we can now suggest that these structures also involve a buckling of the upper nickel layer in a similar manner to that seen with Cu(110).

The Ni(110) $p(4 \times 1)$S structure has been examined using STM[30] and X-ray diffraction.[31] In these studies, the $p(4 \times 1)$ structure was obtained from the

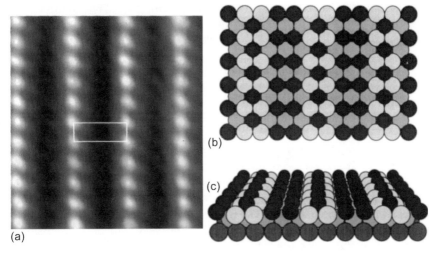

Figure 10.8 p(4 × 1)S Ni(110) surface. (a) High-resolution STM image. (b, c) Top and perspective views, respectively, of model structure. The sulfur atoms are black and the copper atoms are grey with increasing depth indicated by darker colours. (Reproduced from Ref. 30).

reaction of H_2S with both clean and oxidised nickel surfaces. In the latter case there was no change in surface area on conversion of the p(2 × 1)O adlayer, and from this similar nickel atom densities (0.5 ML) were deduced and a structural model involving a buckled surface layer constructed (Figure 10.8). The proposed model involves an S–S distance of only 0.25–3.0 nm, considerably reduced from the 0.38 nm of the p(3 × 2) structure or the 0.33 nm seen in most nickel sulfides. This model is hampered by the constraint that the sulfur adatoms are in existing four-fold sites. In the light of the information from the Cu(110) surface, where the upper copper layer is no longer in registry with the bulk structure, it is clear that sulfur creates its own sites on these surfaces and this model needs further consideration.

10.2.3 Gold and Silver

Despite its otherwise noble character, gold has a very strong affinity towards sulfur, reacting readily with alkanethiols,[32] hydrogen sulfide[33] and, to a lesser extent, sulfur dioxide.[34,35] This specificity towards sulfur makes gold an ideal substrate for the self-assembly of ordered layers using thiol head groups, and there has been extensive work in this area. However, much of this work has concentrated on adsorption from solution. Most work on sulfur adsorption from the gas phase emphasised the Au(111) surface, which, when clean, exhibits a well-known "herringbone" reconstruction. This is a result of stress produced by a 4% increase in gold atoms in the surface compared with the ideal Au(111) plane expected from the bulk structure. It consists of domains of gold atoms sited in fcc and hcp sites separated by domain walls in which gold atoms are in bridge sites. The surface stress is further relieved by partial dislocations in the domain walls, giving rise to "elbow" bends in the domain walls. The latter show an increased contrast in STM images, giving the characteristic zig-zag bright lines seen in STM images of the clean surface and illustrated in Figure 10.9a.

The adsorption of sulfur on the herringbone structure leads to a lifting of the reconstruction and an increase in the roughness of step-edges. Min et al.[34] attributed these changes to the elimination of gold atoms from the surface layer. The STM images show this occurring at low sulfur coverages, resulting in a breakdown in the order of the herringbone reconstruction and, at higher coverages, in its complete removal to give a simple $(\sqrt{3} \times \sqrt{3})$ $R30°$ sulfur LEED pattern. Interestingly, the latter structure was not observed in the STM images, which showed a (1×1) surface. Min et al.[34] attributed this to a mobile sulfur species; a similar conclusion was reached in the case of sulfur on Cu(110).[19] The effect of sulfur adsorption contrasts with oxygen adsorption at the same surface; whereas sulfur lifts the whole herringbone reconstruction, oxygen only removes the "elbows" and not the separate domain structures. This is an indication of the stronger interaction of the sulfur with the surface than oxygen.

The structure of higher sulfur concentrations on the Au(111) surface have been the subject of considerable discussion. STM images of sulfur and alkanethiols adsorbed from solution typically show square-like structures, which Vericat et al.[36] modelled as a result of S_8 clusters on the surface (Figure 10.10). However, using STM to follow the reconstruction of the surface as sulfur adsorbs, Biener et al.[37] reached a different conclusion. They argue that the pits that they observe developing in the gold terraces at high sulfur coverages indicate the incorporation of gold atoms into the surface structure, thus ruling out a simple adsorbate lattice. Their proposal is for a well-defined AuS 2D phase. Figure 10.11 shows the LEED pattern and STM images that support their case for the AuS phase.

In contrast to gold, there has been surprisingly little attention given to sulfur adsorption at silver surfaces; Oudar and co-workers reported[14] adsorption isotherms for sulfur at the three silver basal planes and LEED patterns for some of these coverages. Rovida and Pratesi[38] confirmed these patterns for the Ag(111) surface and Sotto and Boulliard[39] for Ag(100), but there have been no subsequent studies and, in particular, no detailed models produced which might show the extent to which sulfur reconstructs the silver surface.

10.2.4 Platinum, Rhodium, Ruthenium and Rhenium

The adsorption of sulfur at platinum,[40] rhodium,[41] rhenium[42] and ruthenium[43] has been studied predominantly at fcc(111) and hcp(0001) surfaces and shows many similar characteristics. Adsorption is initially into fcc hollow sites of the fcc metals and hcp sites of the hcp metals; at higher coverages, mixed site occupancy occurs. A (2×2) structure is the first to be recorded appearing in the

Studies of Sulfur and Thiols at Metal Surfaces

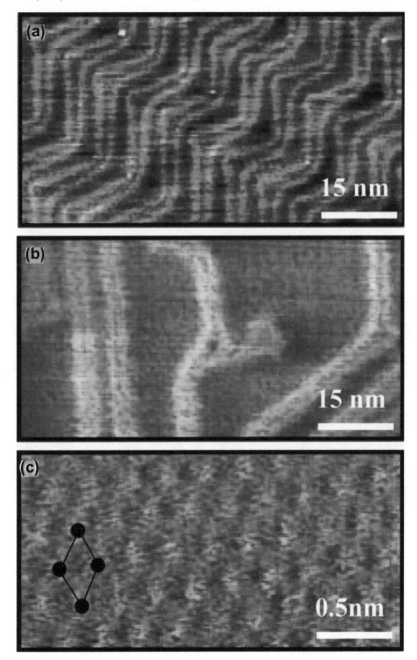

Figure 10.9 STM images showing structural changes induced by sulfur adsorption. (a) Clean Au(111) surface showing very regular herringbone pattern. (b) Close-up of the disordered herringbone pattern at low coverage of sulfur (≤ 0.1 ML). (c) Atomically resolved images of the Au atoms underlying approximately 0.3 ML sulfur adsorbed Au(111). (Reproduced from Ref. 34).

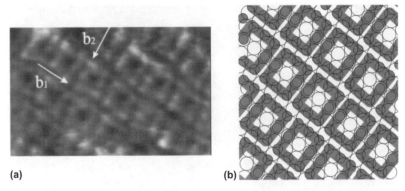

Figure 10.10 (a) STM image (5.2 × 3.2 nm) of the Au(111) surface covered by S_8 surface structures at $E = -0.6$ V in 0.1 M NaOH + 3×10^{-3} M Na_2S. (b) Scheme showing the rectangular S structures (in grey) on Au(111). Large and small circles represent the Au atoms and S atoms, respectively. (Reproduced from Refs. 36 and 37).

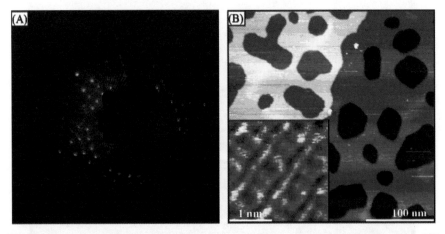

Figure 10.11 A well-ordered 2D AuS phase develops during annealing to 450 K. (A) The structure exhibits a very complex LEED pattern, which can be explained by an incommensurate structure with a nearly quadratic unit cell. (B) STM reveals the formation of large vacancy islands by Oswald ripening which cover about 50% of the surface, thus indicating the incorporation of 0.5 ML of Au atoms into the 2D AuS phase. The 2D AuS phase exhibits a quasi-rectangular structure (inset) and uniformly covers both vacancy islands and terrace areas. (Reproduced from Ref. 37).

LEED patterns at $\theta = 1/4$, but was identified at lower coverages in islands surrounded by mobile sulfur atoms at platinum, rhodium and rhenium surfaces. Sautet and co-workers[42] have analysed the statistical correlations between the intensities of sulfur features in p(2 × 2) islands on rhenium surfaces and also of streaks in areas between islands, which they attribute to sulfur atoms diffusing under the tip (Figure 10.12).

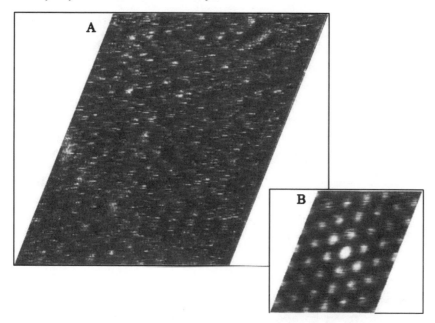

Figure 10.12 (A) 7×7 nm STM image of diffusing sulfur atoms on a Rh(111) surface. The elongation of streaks across the image is due to the STM tip scanning faster across the image than up and down. (B) Correlation image corresponding to (A) showing that the diffusing sulfur is a lattice gas that maintains a local p(2×2) order over several lattice distances. (Reproduced from Ref. 42).

From the size of the streaks in the STM images, a diffusion energy barrier of 0.79 ± 0.1 eV was calculated. With increasing concentration of sulfur, a series of structures develop starting with $(\sqrt{3} \times \sqrt{3})$ $R30°$ at $\theta_S = 1/3$, which is also the most stable thermally of the adsorbed sulfur concentrations. This is followed by a c($\sqrt{3} \times 7$) rect structure, which is the final stage on Pt(111) but which Yoon et al.[41] identify as an intermediate stage in the transformation of the $(\sqrt{3} \times \sqrt{3})$ $R30°$ to a c(4×2) structure on rhodium. It consists of domains separated by evenly spaced "superdense" boundaries in which sulfur is adsorbed at both hcp and fcc sites. The c(4×2) structure corresponding to a coverage of 0.5 contains sulfur atoms in both fcc and hcp hollow sites. On heating, the Rh–c(4×2)–S transforms to a simple (4×4)-S structure. The various transformations are illustrated in Figure 10.13.

10.2.5 Alloy Systems

Bimetallic surfaces are well known for showing radically different chemistry from the individual components and the catalysis industry frequently makes use of these properties to "tune" catalysts, a recent example is the alloying of

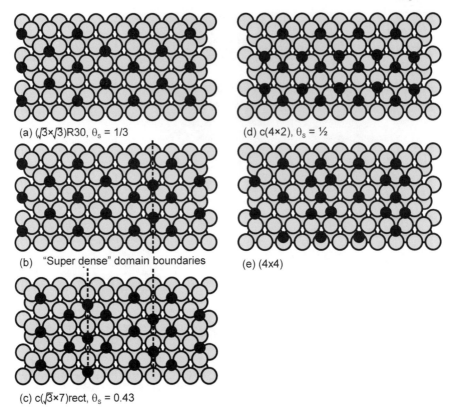

Figure 10.13 Sulfur structures on (111) and (0001) planes of Pt, Rh, Re and Ru. (a) ($\sqrt{3} \times \sqrt{3}$) R30°, $\theta_S = 1/3$; (b) ($\sqrt{3} \times \sqrt{3}$) R30° with a superdense domain boundary; (c) c($\sqrt{3} \times 7$) rect, $\theta_S = 0.43$; (d) c(4 × 2), $\theta_S = 1/2$; (e) (4 × 4). (Adapted from Ref. 40).

gold and palladium to create an effective catalyst for the direct synthesis of H_2O_2.[44] Alloys are used in reforming catalysts where sulfur poisoning is a major problem and also in hydrodesulfurisation catalysts. This interest has resulted in many investigations into sulfur adsorption at bimetallic surfaces, with the majority of work involving bimetallic surfaces created by the adsorption of one component on the well-characterised crystal surface of a second.

The field has been reviewed recently by Rodriguez.[45] STM is ideally suited to the investigation of the complex structures that can occur in these systems and has often been combined with TPD and XPS to provide structural information that complements the surface chemical information. Rodriguez identifies four typical responses to the adsorption of sulfur at bimetallic surfaces. (i) Repulsive interactions between the sulfur and one component are typified by the behaviour of gold adsorbed at the surfaces of rhodium, molybdenum and platinum. Figure 10.14 shows the case[45] of gold at a Ru(0001) surface, where 5% of a monolayer of sulfur results in a dramatic change in the extent of gold island

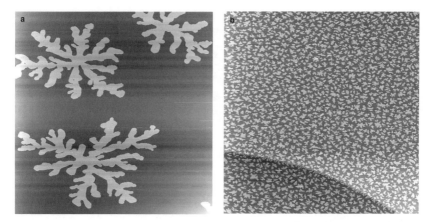

Figure 10.14 STM images (1000 nm) of Au on (a) clean Ru(0001) and (b) Ru(0001) surface with 0.05 ML of preadsorbed sulfur. (Adapted from Ref. 45).

growth due to a reduction in the average diffusion distance of gold adatoms. (ii) Sulfide layer formation: when present as alloy components, copper, silver, molybdenum and ruthenium form sulfide layers on top of the second component. In the case of copper at an Ru(0001) surface,[46] the adsorption of sulfur on a strained double monolayer of copper results in self-organising structures, which, at specific coverages, are beautifully regular and illustrate the large changes in morphology that sulfur can produce (Figure 10.15). The other two are (iii) enhancement of reactivity towards sulfur and (iv) reduced reactivity towards sulfur.

10.3 Sulfur-containing Molecules

Interest in the adsorption of sulfur-containing molecules at metal surfaces been stimulated by a desire to elucidate the decomposition mechanisms of thiols during the catalytic removal of sulfur from feedstocks and the position of thiols as the favoured head groups for adsorbates used to construct self-assembled monolayers. We shall not survey the extensive self-assembled film literature but restrict our discussion to the simpler thiols.

The smallest molecule in the class is methanethiol, which has been studied at several surfaces as an indicator for the behaviour of larger thiols. On all metals, except gold, adsorption results in the dissociation of the S–H bond. Yates and co-workers suggested[47] that the S–H bond of methanethiol is also stable on silver, dissociation being catalysed by adsorbed sulfur present at defect sites. However, this contradicts Jaffey and Madix's results,[56] which showed S–H bond cleavage below 350 K in their study of ethanethiol on Ag(111).

With copper, there is evidence for significant reconstruction occurring on adsorption of thiolates. At Cu(111), for example, two phases, a "pseudo-square"

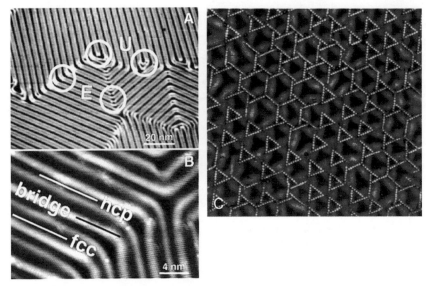

Figure 10.15 Three domains of an anisotropically relaxed second Cu layer striped phase adsorbed on Ru(0001). (A) Large-scale image with the dislocation edges at the domain wall labelled E (elbows) and U (U-turns). (B) Higher-resolution image showing the alternating arrangement of Cu adatoms at different adsorption sites. (C) 7.3 × 6.9 nm STM image of sulfur self-organised in hexagons and equilateral triangles made of 18 sulfur adatoms. At room temperature and fixed S/Cu stoichiometry (0.03 ML for this image) the observed structural patterns fluctuate for hours. (Reproduced from Ref. 46).

and "honeycomb", have been imaged[48] by STM, confirming previous studies by X-ray standing wave and surface extended X-ray absorption fine structure (SEXAFS). What is particularly interesting in this case is that the same structures have been observed when octanethiol is adsorbed at the surface. Figure 10.16 shows STM images of the latter together with proposed structures for the adlayers in each case. The similarity in behaviour suggests that using methane and ethanethiol as models for the larger molecules is a reasonable approximation.

At Cu(110) surfaces, the large-scale movement of copper on adsorption of a thiol is even more apparent; at room temperature, our STM studies[50,51] showed large-scale step movement with the presence of the thiol favouring shorter terraces 2–4 nm in width (Figure 10.17). When the surface is heated to above 450 K, the thiolate decomposes to give a sulfur adlayer with the desorption of carbon in the form of methane, ethane and low concentrations of ethene.[52] The STM results show that the decomposition leaves a surface with sulfur-covered terraces that have extended to 10–20 nm. It has been suggested that the unusual C–H bond scission implicit in the formation of ethene occurs during the large-scale reconstruction.

Studies of Sulfur and Thiols at Metal Surfaces

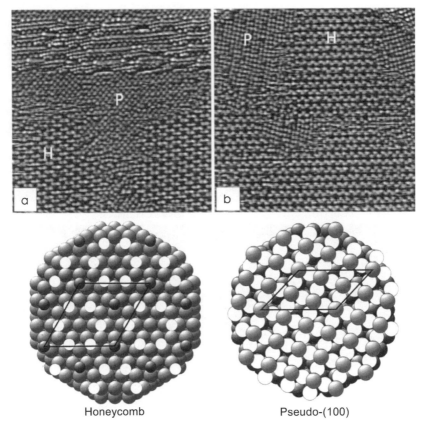

Figure 10.16 20 × 20 nm STM images of the pseudo-square and honeycomb phases of 1-octanethiol on Cu(111). Both images high-pass filtered. The top third of (a) shows some resolution on a multiple step. "H" and "P" indicate regions of the honeycomb and pseudo-(100) reconstructed phases, respectively. Below the STM mages are schematic (plan view) diagrams of the structural models proposed for the honeycomb and pseudo-(100) reconstructions. In both cases the surface unit mesh is shown superimposed on the structure. (Reproduced from Ref. 49).

Methanethiol has been studied at an Ag(111) surface with STM; a $(\sqrt{7} \times \sqrt{7})$ $R19°$ structure was observed[53] that is similar to that seen with sulfur on its own at this surface. However, decomposition of the thiol to sulfur has been carefully ruled out and this conclusion is supported by the STM studies[54] of dimethyl sulfide (DMS) adsorption, which confirmed the $(\sqrt{7} \times \sqrt{7})$ $R19°$ structure. The STM images also show rapid changes in the height profiles of the terraces during the formation of the adsorbed thiolate adlayer (Figure 10.18). The images show some movement and roughening of the steps as a result of the surface thiolate phase formation, but the most significant effect is the appearance of bright regions, which were shown to be approximately 8 Å higher than the terrace on which they develop (approximately three interatomic layer

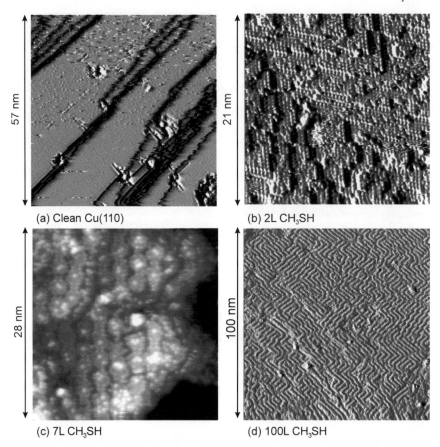

(a) Clean Cu(110) (b) 2L CH$_3$SH
(c) 7L CH$_3$SH   (d) 100L CH$_3$SH

Figure 10.17 STM images of the changes in surface structure observed when methanethiol is adsorbed at a Cu(110) surface at room temperature. (a) Clean surface with terraces approximately 10 nm wide separated by multiple steps. (b) After exposure to 2 L of methanethiol there has been considerable step-edge movement. On the terraces a local c(2 × 2) structure is evident. (c) After a further 7 L exposure, a view of a different area of the crystal shows rounded short terraces; these still retain the c(2 × 2) local structure. (d) After 60 L gross changes to the surface are evident and the STM is unable to image at high resolution.

spacings). By 40 s later, image (d), the islands have been re-accommodated into the terraces. These large-scale movements have been explained by a significant decrease in the density of silver atoms in the silver surface layer.

Various phases have been described[55] for thiolates adsorbed at Au(111) surfaces starting from $(\sqrt{3} \times \sqrt{3})$ $R30°$ at low coverage and including the $3 \times 2\sqrt{3}$, 3×4 and p $\times \sqrt{3}$. All of these are commensurate with the Au(111) surface. In sharp contrast, with Ag(111) an incommensurate $(\sqrt{7} \times \sqrt{7})$ $R19.1°$ structure forms for carbon chains longer than 2. The deviation from commensurate behaviour is thought to be due to repulsive interactions between the close-packed alkyl chains and the reduction in strength of the Ag–Ag bonds to

Studies of Sulfur and Thiols at Metal Surfaces 199

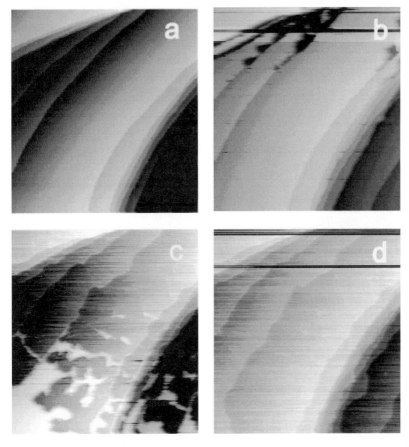

Figure 10.18 Low-magnification STM images (300 × 300 nm) of a stepped area of an Ag(111) surface before and during exposure to dimethyl disulfide. (a) Clean; (b–d) during dosing with DMS at a nominal pressure of 10^{-7} mbar and intervals of approximately 40 s. (Reproduced from Ref. 54).

the bulk compared with that within the surface layer. The latter reduces the energy for deformation and surface mobility.

10.4 Summary

There has been sustained interest in the behaviour of sulfur at surfaces over several decades and a huge literature base exists. The early studies by LEED quickly established a picture of the range of structures that sulfur created at different surfaces and most of these have since been verified with more advanced techniques. However, over the last 20 years our understanding of the nature of these structures has changed significantly. Very few have turned out to be simple overlayer structures, the majority involving surface reconstruction. The major message that emerges from studying sulfur is its ability to create its own adsorption site and the flexibility of the surface in the presence of sulfur. A variety of structural techniques have contributed to our improved understanding of these systems and while STM has certainly played a role, perhaps its major contribution has been to the study of structural changes in real time, revealing details of mass transfer that have elucidated the chemistry of the surface.

References

1. F. S. Karn, J. F. Shultz, R. E. Kelly and R. B. Anderson, *Ind. Eng. Chem. Prod. Res. Dev.*, 1963, **2**, 43.
2. G. J. Hutchings, F. King, I. P. Okoye, M. B. Padley and C. H. Rochester, *J. Catal.*, 1994, **148**, 453–463.
3. J. L. Domange and J. Oudar, *Surf. Sci.*, 1968, **11**, 124; *SCI Monogr.*, 1968, No. 28, 199.
4. L. Ruan, F. Besenbacher, I. Stensgaard and E. Laegsgaard, *Phys. Rev. Lett.*, 1992, **69**, 3523.
5. M. Foss, R. Feidenhansl, M. Nielsen, E. Findeisen, T. Buslaps and R. L. Johnson, *Surf. Sci.*, 1997, **388**, 5.
6. G. J. Jackson, S. M. Driver, D. P. Woodruff, B. C. C. Cowie and R. G. Jones, *Surf. Sci.*, 2000, **453**, 183.
7. S. M. Driver and D. P. Woodruff, *Surf. Sci.*, 2001, **479**, 1.
8. A. E. S. Vonwittenau, Z. Hussain, L. Q. Wang, Z. Q. Huang, Z. G. Ji and D. A. Shirley, *Phys. Rev. B*, 1992, **45**, 13614.
9. E. Vlieg, I. K. Robinson and R. McGrath, *Phys. Rev. B*, 1990, **41**, 7896.
10. H. C. Zeng, R. A. McFarlane and K. A. R. Mitchell, *Phys. Rev. B*, 1989, **39**, 8000.
11. R. McGrath, A. A. Macdowell, T. Hashizume, F. Sette and P. H. Citrin, *Phys. Rev. Lett.*, 1990, **64**, 575.
12. M. L. Colaianni and I. Chorkendorff, *Phys. Rev. B*, 1994, **50**, 8798.
13. M. L. Colaianni, P. Syhler and I. Chorkendorff, *Phys. Rev. B*, 1995, **52**, 2076.
14. M. Kostelitz and J. Oudar, *Surf. Sci.*, 1971, **27**, 176.
15. V. Maurice, J. Oudar and M. Huber, *Surf. Sci.*, 1987, **187**, 312.
16. T. M. Parker, N. G. Condon, R. Lindsay, G. Thornton and F. M. Leibsle, *Surf. Rev. Lett.*, 1994, **1**, 705.
17. I. Stensgaard, L. Ruan, F. Besenbacher, F. Jensen and E. Laegsgaard, *Surf. Sci.*, 1992, **270**, 81.
18. A. Atrei, A. L. Johnson and D. A. King, *Surf. Sci.*, 1991, **254**, 65.
19. A. F. Carley, P. R. Davies, R. V. Jones, K. R. Harikumar, G. U. Kulkarni and M. W. Roberts, *Surf. Sci.*, 2000, **447**, 39.
20. M. Perdereau and J. Oudar, *Surf. Sci.*, 1970, **20**, 80.
21. T. Edmonds, J. McCarrol and R. C. Pitkethly, *J. Vac. Sci. Technol. A*, 1971, **8**, 68.

22. D. R. Warburton, P. L. Wincott, G. Thornton, F. M. Quinn and D. Norman, *Surf. Sci.*, 1989, **211**, 71.
23. L. Ruan, I. Stensgaard, F. Besenbacher and E. Laegsgaard, *J. Vac. Sci. Technol. B*, 1994, **12**, 1772.
24. D. P. Woodruff, *Phys. Rev. Lett.*, 1994, **72**, 2499.
25. M. Foss, R. Feidenhansl, M. Nielsen, E. Findeisen, R. L. Johnson, T. Buslaps, I. Stensgaard and F. Besenbacher, *Phys. Rev. B*, 1994, **50**, 8950.
26. A. Partridge, G. J. Tatlock and F. M. Leibsle, *Surf. Sci.*, 1997, **381**, 92.
27. R. J. Madix, S. B. Lee and M. Thornburg, *J. Vac. Sci. Technol. A*, 1983, **1**, 1254.
28. A. R. Alemozafar and R. J. Madix, *J. Phys. Chem. B*, 2005, **109**, 11307.
29. D. R. Warburton, G. Thornton, D. Norman, C. H. Richardson, R. McGrath and F. Sette, *Surf. Sci.*, 1987, **189**, 495.
30. L. Ruan, I. Stensgaard, E. Laegsgaard and F. Besenbacher, *Surf. Sci.*, 1993, **296**, 275.
31. M. Foss, R. Feidenhansl, M. Nielsen, E. Findeisen, T. Buslaps, R. L. Johnson, F. Besenbacher and I. Stensgaard, *Surf. Sci.*, 1993, **296**, 283.
32. C. Vericat, M. E. Vela, G. A. Benitez, J. A. M. Gago, X. Torrelles and R. C. Salvarezza, *J. Phys.: Condens. Matter.*, 2006, **18**, R867.
33. I. Touzov and C. B. Gorman, *Langmuir*, 1997, **13**, 4850.
34. B. K. Min, A. R. Alemozafar, M. M. Biener, J. Biener and C. M. Friend, *Top. Catal.*, 2005, **36**, 77.
35. G. Liu, J. A. Rodriguez, J. Dvorak, J. Hrbek and T. Jirsak, *Surf. Sci.*, 2002, **505**, 295.
36. C. Vericat, G. Andreasen, M. E. Vela and R. C. Salvarezza, *J. Phys. Chem. B*, 2000, **104**, 302.
37. M. M. Biener, J. Biener and C. M. Friend, *Langmuir*, 2005, **21**, 1668.
38. G. Rovida and F. Pratesi, *Surf. Sci.*, 1981, **104**, 609.
39. M. P. Sotto and J. C. Boulliard, *Surf. Sci.*, 1985, **162**, 285.
40. H. A. Yoon, N. Materer, M. Salmeron, M. A. VanHove and G. A. Somorjai, *Surf. Sci.*, 1997, **376**, 254.
41. H. A. Yoon, M. Salmeron and G. A. Somorjai, *Surf. Sci.*, 1998, **395**, 268.
42. J. C. Dunphy, P. Sautet, D. F. Ogletree, O. Dabbousi and M. B. Salmeron, *Phys. Rev. B*, 1993, **47**, 2320.
43. T. Muller, D. Heuer, H. Pfnur and U. Kohler, *Surf. Sci.*, 1996, **347**, 80.
44. M. D. Hughes, Y. J. Xu, P. Jenkins, P. McMorn, P. Landon, D. I. Enache, A. F. Carley, G. A. Attard, G. J. Hutchings, F. King, E. H. Stitt, P. Johnston, K. Griffin and C. J. Kiely, *Nature*, 2005, **437**, 1132.
45. J. A. Rodriguez, *Prog. Surf. Sci.*, 2006, **81**, 141.
46. J. Hrbek, J. de la Figuera, K. Pohl, T. Jirsak, J. A. Rodriguez, A. K. Schmid, N. C. Bartelt and R. Q. Hwang, *J. Phys. Chem. B*, 1999, **103**, 10557.
47. J. G. Lee, J. Lee and J. T. Yates, *J. Phys. Chem. B*, 2004, **108**, 1686.
48. S. M. Driver and D. P. Woodruff, *Surf. Sci.*, 2000, **457**, 11.
49. S. M. Driver and D. P. Woodruff, *Langmuir*, 2000, **16**, 6693.

50. A. F. Carley, P. R. Davies, R. V. Jones, K. R. Harikumar, M. W. Roberts and C. J. Welsby, *Top. Catal.*, 2003, **22**, 161.
51. A. F. Carley, P. R. Davies, R. V. Jones, K. R. Harikumar and M. W. Roberts, *Surf. Sci.*, 2001, **490**, L585.
52. Y. H. Lai, C. T. Yeh, S. H. Cheng, P. Liao and W. H. Hung, *J. Phys. Chem. B*, 2002, **106**, 5438.
53. A. L. Harris, L. Rothberg, L. H. Dubois, N. J. Levinos and L. Dhar, *Phys. Rev. Lett.*, 1990, **64**, 2086.
54. M. Yu, S. M. Driver and D. P. Woodruff, *Langmuir*, 2005, **21**, 7285.
55. M. Kawasaki and M. Iino, *J. Phys. Chem. B*, 2006, **110**, 21124.
56. D. M. Jaffey and R. J. Madix, *Surf. Sci.*, 1971, **27**, 176.

Further Reading

C. Vericat, M. E. Vela, G. A. Benitez, J. A. M. Gago, X. Torrelles and R. C. Salvarezza, Surface characterisation of sulfur and alkanethiol self-assembled monolayers on Au(111), *J. Phys.: Condens. Matter*, 2006, **18**, R867.

J. A. Rodriguez, The chemical properties of bimetallic surfaces: importance of ensemble and electronic effects in the adsorption of sulfur and SO_2, *Prog. Surf. Sci.*, 2006, **81**, 141.

F. Schreiber, Structure and growth of self-assembling monolayers, *Prog. Surf. Sci.*, 2000, **65**, 151.

CHAPTER 11
Surface Engineering at the Nanoscale

"For the world was built in order and the atoms march in tune"

Ralph Waldo Emerson

11.1 Introduction

Up until the last few decades, surface engineering at the microscopic level was largely chemical in nature, using reactants to pacify or activate one surface towards another; topographical changes were only possible at the macroscopic level, although annealing, polishing, *etc.*, did have microscopic consequences. The ability to study topography in the nano-domain has led to a better understanding of the role played by the structure at this scale; take, for example, the super-hydrophobic properties of the lotus leaf, which are partly due to its nanoscale structure and which have prompted considerable research by technologists wishing to reproduce these effects on other surfaces. Another example is the wing surface of *Cicada orni*,[1] whose nanostructured wing scales prevent the accumulation of dust particles (Figure 11.1).

Electron lithography has proven be an enormously successful tool for the engineering of surface topography at resolutions which are now approaching 50 nm. However, while the size of structures that can be created by lithographical means has continued to decrease, there is widespread recognition that this "top-down" approach is nearing its limit. Single atom positioning[2,3] with an STM tip has shown what is ultimately possible for surface patterning; structures such as commercial logos and atomic-scale electron "corrals" have been produced (Figure 11.2). However, this linear atom by atom technique cannot produce the large surface areas necessary for most technological applications; for these, parallel construction techniques such as self-assembly are needed.

Self-assembled monolayers of amphiphilic molecules have been deposited at surfaces since Langmuir and Blodgett developed their dip coating deposition method in 1937.[4] These were briefly discussed in Chapter 10 in relation to thiol

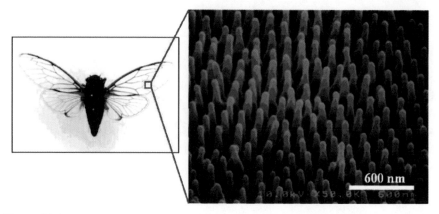

Figure 11.1 (a) Cicada orni, which avoids dust particles accumulating on its wings by virtue of its nanostructure; (b) FE-SEM micrograph of the wing surface showing the regularly aligned nanoposts which minimize interactions with the dust particles. (Reproduced from Ref. 1).

adsorption, since these have been the mainstay of the assembly of molecules from gas and liquid phases at gold surfaces. In this case, the surface layer is characterised by close-packed molecules which can be deposited monolayer by monolayer. While this permits very precise control over the structure of the adsorbates perpendicular to the surface, the structure parallel to the surface is harder to direct. Surface chemists have therefore been attempting to organise molecules by taking advantage of attractive intermolecular forces to create hierarchical structures with the potential to host guest molecules or clusters and thereby tailor surfaces to particular functions. Ideas have been adapted from studies of self-assembly in three dimensions,[5,6] but the strong influence of the substrate means that comparisons between 2D and 3D assembly are not always straightforward.

11.2 "Bottom-up" Surface Engineering

A simple example of "bottom-up" lithography, made possible by the availability of techniques that can resolve structures at the nanoscale, is the case of nitride structures at a Cu(100) surface. LEED studies had shown a c(2 × 2) structure for nitride adsorbed at a Cu(100) surface, but STM studies[7] showed that the nitride structure is characterised by square islands with well-defined spaces between the islands. Leibsle and co-workers imaginatively used these natural self-assembled boxes to confine cobalt[7] and iron[8] into small domains, the aim being to construct a surface in which the magnetic properties could be controlled by the surface topography (Figure 11.3).

The drawback of the nitride/Cu(100) approach is its narrow application; while the principle is sound, the nanoscale nitride corrals only work on copper and a limited number of other single-crystal surfaces. For more realistic applications, a more generic approach is needed. The obvious place to start is to make use of intermolecular forces to control the arrangement of molecules

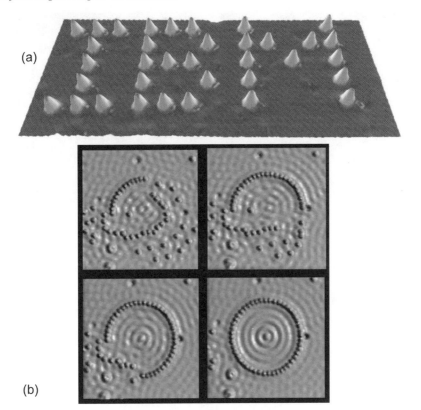

Figure 11.2 (a) Xenon atoms at 4 K, positioned on a nickel(110) surface using an STM tip (Reproduced from Ref. 3). (b) A "quantum corral" built by positioning iron atoms with an STM tip on a Cu(111) surface. (Reproduced from Ref. 2).

at a surface. There have been numerous studies of systems that relate to this topic and the role of intermolecular forces in controlling two-dimensional structures at surfaces has been discussed for many years. However, the advent of scanning probe microscopies has made the study of these systems more accessible, and a plethora of studies of related systems have appeared in recent years. Of these we shall discuss only a few that illustrate some of the general principles that have developed in the field; broadly these will be characterised by the nature of the bonding involved in the adsorbate structures.

11.2.1 Van der Waals Forces

Phthalocyanines have attracted particular attention as potential surface modifiers due to their stability and tendency to form ordered structures directed by dispersion forces. They are inherently host–guest structures with a readily interchangeable coordinating metal ion, which in the solid state results in a "tunable" bandgap. At a surface, in addition to possibly interesting electronic

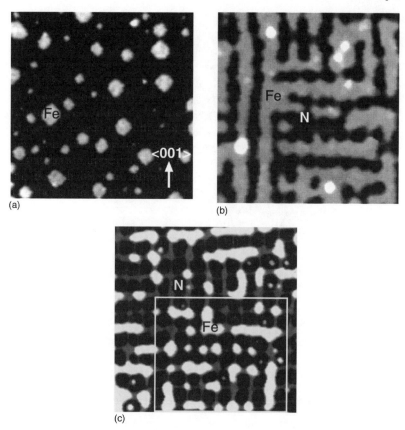

Figure 11.3 A series of three 60 × 60 nm STM images showing how the growth of Fe on Cu(100) surfaces can be controlled by the presence of N islands. (a) Clean Cu(100) surface on which 0.13 ML of Fe was deposited at room temperature. The Fe has grown as small randomly spaced islands, with island edges running in the <011> directions. (b) A Cu(100)–c(2 × 2)N surface on which 0.4 ML of Fe has been grown. The epitaxially grown Fe is channelled into the regions between the N islands forming long narrow stripes with island edges now running in the <001> directions. Some second-layer growth can also be observed, one area of which has been boxed. (c) A Cu(100)–c(2 × 2)N surface on which 0.2 ML of Fe has been grown. Again the epitaxially grown Fe is channelled into regions between the N islands. The region enclosed in the rectangle shows an almost perfect two-dimensional square array of Fe islands on this surface. (Reproduced from Ref. 8).

properties, the interchangeable metal ion has the potential to provide customisable surface chemistry. An early study of phthalocyanine adsorption by Müller in 1950 reported[9] the direct imaging of a phthalocyanine adsorbed on the emitter tip of a field emission microscope. He suggested that four bright spots in the image corresponded to the four π-ring systems of the molecule. However, interpretation of the data was somewhat controversial; a more widely accepted approach published in 1973 by Graham et al.[10] involved a copper phthalocyanine adsorbed on an iridium tip and coated in platinum. The cavities in the platinum created by the adsorbed molecules were imaged using field ion emission. Since these early studies, interest in phthalocyanine adsorption has remained strong, partly because they are useful model systems for haemoglobin and chlorophyll and partly because of their potentially useful electronic properties. Buchholz and Somorjai[11] studied the adsorption of copper and iron phthalocyanines at Cu(111) and Cu(100) surfaces using LEED in 1977 and observed epitaxial growth. From NEXAFS data and molecular orbital calculations, Koch and co-workers[12] reported that nickel phthalocyanine adsorbed with its molecular plane parallel to the substrate. Phthalocyanine adsorption at surfaces was amongst the first systems to be studied by STM; an early investigation by Gimzewski et al. of Cu phthalocyanines at a polycrystalline silver surface[13] resolved individual molecules. Later, in a study of the same molecule at a Cu(100) surface, Lippel et al.[14] observed ordered structures and individual Cu phthalocyanine molecules were imaged together with the underlying lattice, allowing the identification of a precise adsorption site. Many studies have made use of the ability to change the central metal ion to explore imaging characteristics of the adsorbed phthalocyanines[15–19] and more recently Koudia et al.[20] investigated the addition of chlorine atoms to the zinc phthalocyanine structure to introduce hydrogen bonding properties (Figure 11.4). After dosing the modified phthalocyanines, they found a progression from a structure dominated by dispersion forces to one controlled by hydrogen bonding over a period of some 70 h.

11.2.2 Hydrogen Bonding

Hydrogen bonding has many advantages for the self-assembly of structures at surfaces: they are sufficiently strong to control the orientation of a molecule in a structure at temperatures slightly above room temperature, but can be broken and reformed under relatively mild conditions allowing a rapid equilibration of a system to a minimum-energy structure at low temperatures. This allows the annealing out of domain boundaries without recourse to temperatures which might destroy the adsorbed molecules. Similarly, such bonds are relatively flexible, ameliorating the problems of strain that are frequently caused by the development of a structure with a poor lattice fit to the substrate. A good example comes from work by Barth and co-workers,[21] who studied the adsorption of terephthalic acid (TPA) at an Au(111) surface (Figure 11.5). "Head to tail" hydrogen bonding dominated the interactions between the adsorbed molecules, but the authors were able to determine that the average hydrogen bond length is greater than that seen in the solid-state structure.

Furthermore, the Au(111) substrate exhibits a herringbone reconstruction which was not lifted when the TPA was adsorbed. The authors were able to show that the adsorption site remained fixed over this reconstruction, leading to local variations in the hydrogen bond lengths within the adlayer. Polanyi and co-workers[22] recently utilised hydrogen bonding between haloalkanes to generate self-assembled corrals around single atoms at a silicon surface.

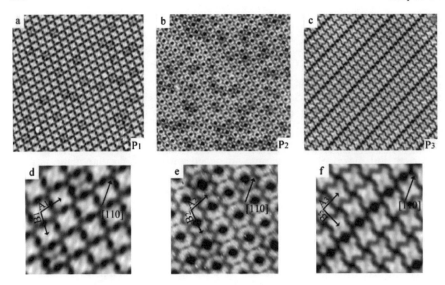

Figure 11.4 STM images obtained at room temperature for the three two-dimensional arrangements of ZnPcCl$_8$ molecules deposited on Ag(111): (a) immediately after the deposit, phase P1; (b) about 40 h after deposit, phase P2; (c) about 70 h after deposit, phase P3; (d–f) zoom on phase P1 (lattice parameters A1 = B1 = 18 Å), intermediate phase P2 (A2 = 15 Å, B2 = 18 Å) and final phase P3 (A3 = 15 Å, B3 = 15 Å), respectively. Dimensions of the upper and lower images: 30 × 30 and 7.5 × 7.5 nm, respectively. (Reproduced from Ref. 20).

Scanning tunnelling spectroscopy suggested that the corrals result in electron transfer to the corralled atom. In a similar fashion, Pennec et al.[23] used hydrogen bonding between methionine molecules adsorbed at a Ag(111) surface to create "supramolecular gratings". The methionine molecules self-assemble into regular one-dimensional domain structures without causing reconstruction of the silver surface. The one-dimensional domains of clean copper between the methionine walls act as traps for surface electron states which can be detected by STM.

Beton and co-workers extended the hydrogen bonding approach to two-component systems, generating a number of structures that utilise different molecular motifs.[24–26] In the case of perylene tetracarboxylic diimide (PTCDI) co-adsorbed with melamine (1,3,5-triazine-2,4,6-triamine) on a silver-terminated silicon surface, a network is formed in which the straight edges correspond to PTCDI with melamine at the vertices (Figure 11.6). The network shows large-area pores that the authors used to trap heptamers of C$_{60}$ molecules.

11.2.3 Chiral Surfaces from Prochiral Adsorbates

One particular example in which "bottom-up" engineering has a direct impact on the properties of a surface is where the adsorbates impart particular

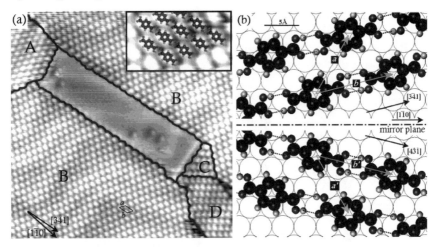

Figure 11.5 (a) Coexistence of different TPA rotational domains on Au(111). Four orientations are present: A, C and D are rotated by 120° relative to each other; B represents the mirror symmetric arrangement of D with respect to [1$\bar{1}$0]. (STM image size 28 × 28 nm). The oval shape of the molecules and the anisotropy of the domain boundaries determine the molecular orientation (inset). (b) Model for the molecular superstructure; a (along the [0$\bar{1}$1] direction) and b (along the [3$\bar{4}$1] direction) are the lattice vectors, a and b are the base vectors for the lattice of opposite chirality (mirror symmetry of a and b with respect to [1$\bar{1}$0]). The hydrogen bonds are indicated by dashed lines. For simplicity the Au(111) substrate is modelled here as perfectly hexagonal. The molecule adsorption site is arbitrary. (Reproduced from Ref. 21).

symmetry properties. In particular, chiral adsorbate modifiers have been shown to convert non-chiral surfaces into enantioselective heterogeneous catalysts; examples include cinchona alkaloids[27] at platinum surfaces and tartaric acid on nickel catalysts. Lambert and co-workers have used STM extensively to study[28–31] the former system. In the case of tartaric acid, Lorenzo and co-workers[32,33] used STM to investigate the chiral surface structures created when (S,S)- or (R,R)-tartaric acid is adsorbed at Cu(110) surfaces. The study showed chiral structures with different "handedness" for the two chiral molecules and suggested that these structures were responsible for the enantioselectivity of the catalysts by creating enantiospecific adsorption sites. However, this area remains controversial, with the question of the size of domain possible at the surface of a nano-sized catalyst particle being central to the debate.

11.2.4 Covalently Bonded Systems

An alternative to using van der Waals forces to organise molecules at surfaces is to covalently bond monomers. Haq and Richardson,[34] for example, have attempted to develop PMDA–ODA oligomers using controlled imide coupling reactions between pyromellitic dianhydride and 4,4-oxydianiline at clean and oxidised Cu(110) surfaces. Reflection–absorption infrared spectroscopy (RAIRS) was used to follow the reactions as a function of temperature and coverage and provided evidence for multilayer growth.

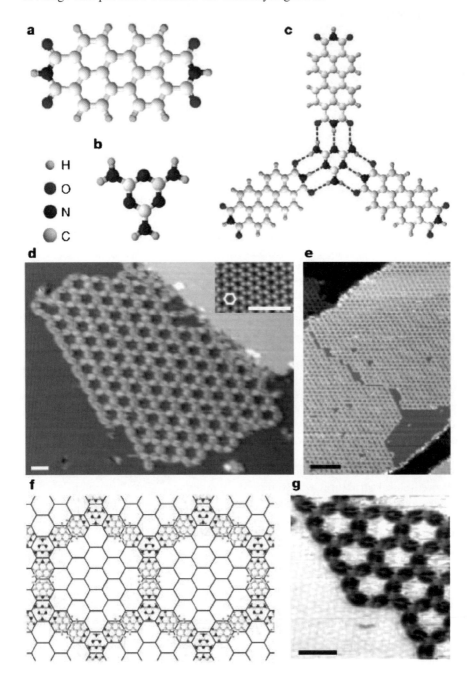

An alternative approach was adopted by Ozaki and co-workers,[35–37] who studied the polymerisation of a monolayer of the dialkyldiacetylene 17, 19- hexatriacontadiyne (HTDY) molecules adsorbed at the surface of highly oriented pyrolytic graphite (Figure 11.7). Exposure of the monolayer to UV light or to low-energy electrons results in two-dimensional polymerisation, forming columns of polydiacetylene and polyacetylene chains alternately cross-linked to the rows of alkyl chains. The authors described the product as a 0.4 nm thick "atomic cloth" and characterised the product using Penning ionisation spectroscopy (PIS). PIS measures electron emission stimulated by the impact of metastable He atoms at the surface; it is sensitive only to the uppermost atomic layer and provides information on the occupied electronic levels. Evidence was presented for extended delocalised structures in the polymerised adlayer; however, the domain size of the polymerised system could not be determined from this method and there is little information as to whether lattice strain between the polymerised structure and the substrate was an issue. More recently, Okawa and Aono[38,39] returned to this system using a voltage pulse from an STM tip to initiate polymerisation within the ordered monolayer of a diacetylene compound, 10,12-nonacosadiynoic acid, adsorbed on a graphite surface They showed that the polymerisation reaction initiated in this way propagated in a linear direction and could be terminated by a carefully positioned defect in the structure Figure (11.7). Since the polymerisation reaction results in a conjugated linear polymer, the authors suggested that the approach might be used as a method of generating well-controlled nanowires between features on a surface.

11.3 Surface Engineering Using Diblock Copolymer Templates

Diblock copolymers consist of contiguous sequences of two different covalently bound monomer units, arranged in an –A-A-A-B-B-B-B- structure. In an appropriate solvent, the diblock copolymers spontaneously self-assemble into micelles with cores which are essentially pure in one component and a diameter

Figure 11.6 Self-assembly of a PTCDI–melamine supramolecular network. (a, b) Chemical structures of PTCDI (a) and melamine (b). (c) Schematic diagram of a PTCDI–melamine junction. Dotted lines represent the stabilizing hydrogen bonds between the molecules. (d) STM image of a PTCDI–melamine network. Inset: high-resolution view of the Ag/Si(111)–($\sqrt{3} \times \sqrt{3}$) R30° substrate surface; the vertices and centres of hexagons correspond, respectively, to the bright (Ag trimers) and dark (Si trimers) topographic features in the STM image (surface lattice constant, a = 0.665 nm[5]). Scale bars, 3 nm. (e) STM image of large-area network, with domains extending across terraces on the Ag/Si(111)–($\sqrt{3} \times \sqrt{3}$) R30° surface. Scale bar, 20 nm. (f) Schematic diagram showing the registry of the network with the surface. (g) Inverted contrast image of the network. Scale bar, 3 nm. (Reproduced from Ref. 24).

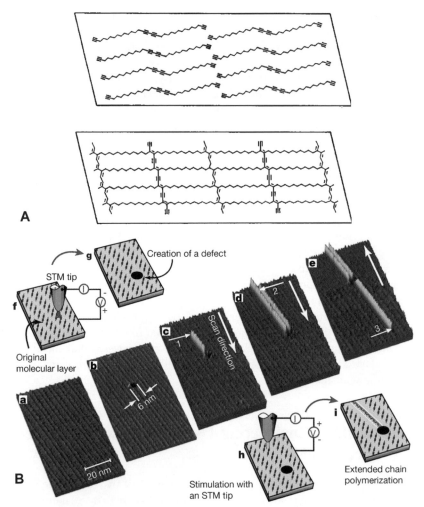

Figure 11.7 (A) Intramonolayer polymerisation of 1,15,17,31-dotriacontatetrayne (DTTY). (a) Arrangement of DTTY molecules in a monolayer. (b) Product of photopolymerisation, atomic cloth: a single sheet of a cloth–like macromolecule comprising the columns of polydiacetylene and polyacetylene chains alternately cross-linked to the rows of alkyl chains (Reproduced from Ref. 24). (B) STM images and diagrams showing the process of controlling the initiation and termination of linear chain polymerisation with an STM tip. (a) STM image of the original monomolecular layer of 10,12-nonacosadiynoic acid. (b) Creation of an artificial defect in advance in the monomolecular layer using the STM tip. (c) First chain polymerisation, initiated at the point indicated by arrow (1) using an STM tip and terminated at the artificial defect. (d) Second chain polymerisation, initiated at arrow (2). (e) Third chain polymerisation, initiated at arrow (3). (f, g) Creation of an artificial defect in advance with an STM tip. (h, i) Initiation of chain polymerisation with an STM tip and termination of the polymerisation at the artificial defect. (Reproduced from Ref. 39).

d governed by the chain dimension of the second component, typically this ranges 10–150 nm. Investigations of the structures of these materials in solution revealed a degree of order and these soft matter phases were exploited by Attard *et al.*[40] and later by Antonietti and co-workers[41] to create ordered three-dimensional silica structures with controlled pore sizes, an approach known as "true liquid crystal templating" or alternatively "nanocasting". An adaptation of this approach, pioneered by Möller and co-workers,[42–44] utilises the ability of the copolymer micelles to adsorb metal salts selectively into the centre of the micelle. The micellular solution is cast on to a flat surface using spin. Since entropic effects prevent the copolymer micelles from coalescing, they retain their spherical structure even after deposition on a surface and form a periodic close-packed hexagonal arrangement with an inter-particle spacing related to the chain length. The organic adlayer surrounding the particles can be selectively removed by oxidation, leaving the metal oxides that had been trapped within the micelle core. These monodispersed oxide particles can be reduced back to nanoparticulate metal by treatment with hydrogen. Surprisingly, in the systems reported very little sintering occurs and extended arrays of metal particles with very well-defined spacings and size distributions have been generated (Figure 11.8a–d). The technique was recently reviewed by Haryono and Binder.[45] More recently, the diblock copolymer deposition method has been developed by Lu and co-workers[46,47] to produce metal catalysts which are active for the growth of carbon nanotubes (Figure 11.8e) and Kielbassa *et al.*[48] have used the diblock copolymer approach to deposit gold particles on TiO_2.

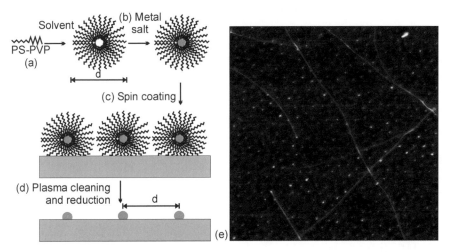

Figure 11.8 Formation of ordered nanoparticles of metal from diblock copolymer micelles. (a) Diblock copolymer; (b) metal salt partition to centres of the polymer micelles; (c) deposition of micelles at a surface; (d) micelle removal and reduction of oxide to metal. (e) AFM image of carbon nanotubes and cobalt catalyst nanoparticles after growth (height scale, 5 nm; scan size, 1 × 1 μm). [Part (e) reproduced from Ref. 47].

11.4 Summary

The self-assembly of molecules at surfaces to create purpose-designed nanoscale structures has advanced significantly since the advent of proximal probe methods. So far, experiments have established the feasibility of using van der Waals forces to control two-dimensional topography at the nanoscale and experiments have begun to make use of this technique to create templates for the organisation of other molecules/nanoparticles. The next step is significant; exploiting these structures to control the development of technologically useful structures on a microscale as has been achieved in three dimensions.

References

1. W. Lee, M. K. Jin, W. C. Yoo and J. K. Lee, *Langmuir*, 2004, **20**, 7665.
2. M. F. Crommie, C. P. Lutz and D. M. Eigler, *Science*, 1993, **262**, 218.
3. D. M. Eigler and E. K. Schweizer, *Nature*, 1990, **344**, 524.
4. K. B. Blodgett and I. Langmuir, *Phys. Rev.*, 1937, **51**, 964.
5. L. J. Prins, D. N. Reinhoudt and P. Timmerman, *Angew. Chem. Int. Ed.*, 2001, **40**, 2383.
6. M. J. Krische and J. M. Lehn, in *Molecular Self-assembly*, ed. D. M. P. Mingos, Springer, Heidelberg, 2000, **196**, p. 3.
7. S. L. Silva, C. R. Jenkins, S. M. York and F. M. Leibsle, *Appl. Phys. Lett.*, 2000, **76**, 1128.
8. T. M. Parker, L. K. Wilson, N. G. Condon and F. M. Leibsle, *Phys. Rev. B*, 1997, **56**, 6458.
9. E. W. Müller, *Z. Naturforsch. Teil A*, 1950, **5**, 473.
10. W. R. Graham, F. Hutchinson and D. A. Reed, *J. Appl. Phys.*, 1973, **44**, 5155.
11. J. C. Buchholz and G. A. Somorjai, *J. Chem. Phys.*, 1977, **66**, 573.
12. M. L. M. Rocco, K. H. Frank, P. Yannoulis and E. E. Koch, *J. Chem. Phys.*, 1990, **93**, 6859.
13. J. K. Gimzewski, E. Stoll and R. R. Schlittler, *Surf. Sci.*, 1987, **181**, 267.
14. P. H. Lippel, R. J. Wilson, M. D. Miller, C. Woll and S. Chiang, *Phys. Rev. Lett.*, 1989, **62**, 171.
15. X. Lu and K. W. Hipps, *J. Phys. Chem. B*, 1997, **101**, 5391.
16. C. Seidel, C. Awater, X. D. Liu, R. Ellerbrake and H. Fuchs, *Surf. Sci.*, 1997, **371**, 123.
17. X. Lu, K. W. Hipps, X. D. Wang and U. Mazur, *J. Am. Chem. Soc.*, 1996, **118**, 7197.
18. D. E. Barlow and K. W. Hipps, *J. Phys. Chem. B*, 2000, **104**, 5993.
19. F. J. Williams, O. P. H. Vaughan, K. J. Knox, N. Bampos and R. M. Lambert, *Chem. Commun.*, 2004, 1688.
20. M. Koudia, M. Abel, C. Maurel, A. Bliek, D. Catalin, M. Mossoyan, J. C. Mossoyan and L. Porte, *J. Phys. Chem. B*, 2006, **110**, 10058.

21. S. Clair, S. Pons, A. P. Seitsonen, H. Brune, K. Kern and J. V. Barth, *J. Phys. Chem. B*, 2004, **108**, 14585.
22. S. Dobrin, K. R. Harikumar, R. V. Jones, N. Li, I. R. McNab, J. C. Polanyi, P. A. Sloan, Z. Waqar, J. Yang, S. Ayissi and W. A. Hofer, *Surf. Sci.*, 2006, **600**, L43.
23. W. A. Y. Pennec, A. Schiffrin, A. Weber-Bargioni, A. Riemann and J. V. Barth, *Nat. Nanotechnol.*, 2007, **2**, 99.
24. J. A. Theobald, N. S. Oxtoby, M. A. Phillips, N. R. Champness and P. H. Beton, *Nature*, 2003, **424**, 1029.
25. L. M. A. Perdigao, N. R. Champness and P. H. Beton, *Chem. Commun.*, 2006, 538.
26. L. M. A. Perdigao, E. W. Perkins, J. Ma, P. A. Staniec, B. L. Rogers, N. R. Champness and P. H. Beton, *J. Phys. Chem. B*, 2006, **110**, 12539.
27. P. B. Wells, in *Surface Chemistry and Catalysis*, ed. A. F. Carley, P. R. Davies, G. J. Hutchings and M. S. Spencer, Kluwer, New York, 2002, p. 295.
28. M. J. Stephenson and R. M. Lambert, *J. Phys. Chem. B*, 2001, **105**, 12832.
29. J. M. Bonello, R. Lindsay, A. K. Santra and R. M. Lambert, *J. Phys. Chem. B*, 2002, **106**, 2672.
30. J. M. Bonello, F. J. Williams and R. M. Lambert, *J. Am. Chem. Soc.*, 2003, **125**, 2723.
31. A. I. McIntosh, D. J. Watson, J. W. Burton and R. M. Lambert, *J. Am. Chem. Soc.*, 2006, **128**, 7329.
32. M. O. Lorenzo, C. J. Baddeley, C. Muryn and R. Raval, *Nature*, 2000, **404**, 376–379.
33. M. O. Lorenzo, S. Haq, T. Bertrams, P. Murray, R. Raval and C. J. Baddeley, *J. Phys. Chem. B*, 1999, **103**, 10661–10669.
34. S. Haq and N. V. Richardson, *J. Phys. Chem. B*, 1999, **103**, 5256.
35. H. Ozaki, T. Funaki, Y. Mazaki, S. Masuda and Y. Harada, *J. Am. Chem. Soc.*, 1995, **117**, 5596.
36. H. Ozaki, M. Kasuga, T. Tsuchiya, T. Funaki, Y. Mazaki, M. Aoki, S. Masuda and Y. Harada, *J. Chem. Phys.*, 1995, **103**, 1226.
37. H. Ozaki, *J. Electron. Spectrosc. Relat. Phenom.*, 1995, **76**, 377.
38. Y. Okawa and M. Aono, *Top. Catal.*, 2002, **19**, 187.
39. Y. Okawa and M. Aono, *Nature*, 2001, **409**, 683.
40. G. S. Attard, J. C. Glyde and C. G. Goltner, *Nature*, 1995, **378**, 366.
41. C. G. Goltner, S. Henke, M. C. Weissenberger and M. Antonietti, *Angew. Chem. Int. Ed.*, 1998, **37**, 613.
42. J. P. Spatz, P. Eibeck, S. Mossmer, M. Möller, E. Y. Kramarenko, P. G. Khalatur, Potemkin, II, A. R. Khokhlov, R. G. Winkler and P. Reineker, *Macromolecules*, 2000, **33**, 150.
43. J. P. Spatz, S. Mossmer and M. Möller, *Chem. Eur. J.*, 1996, **2**, 1552.
44. R. Glass, M. Möller and J. P. Spatz, *Nanotechnology*, 2003, **14**, 1153.
45. A. Haryono and W. H. Binder, *Small*, 2006, **2**, 600.
46. J. Q. Lu and S. S. Yi, *Langmuir*, 2006, **22**, 3951.

47. J. Lu, S. S. Yi, T. Kopley, C. Qian, J. Liu and E. Gulari, *J. Phys. Chem. B*, 2006, **110**, 6655.
48. S. Kielbassa, A. Habich, J. Schnaidt, J. Bansmann, F. Weigl, H. G. Boyen, P. Ziemann and R. J. Behm, *Langmuir*, 2006, **22**, 7873.

Further Reading

X. J. Feng and L. Jiang, Design and creation of superwetting/antiwetting surfaces, *Adv. Mater.*, 2006, **18**, 3063.

J. V. Barth, G. Costantini and K. Kern, Engineering atomic and molecular nanostructures at surfaces, *Nature*, 2005, **437**, 671.

A. Haryono and W. H. Binder, Controlled arrangement of nanoparticle arrays in block-copolymer domains, *Small*, 2006, **2**, 600.

Epilogue

"Chemistry without catalysis would be like a sword without a handle, a light without brilliance, a bell without sound"

A. Mittasch

In 1990, one of us, in an address at the British Association Meeting, concluded the talk on "Catalysis" with the cartoon shown below depicting one of the major problems that needed to be addressed in order to limit the Greenhouse Phenomenon. Could a catalyst be designed so that the emission of the greenhouse gas, CO_2, from the vehicle exhaust systems be eliminated by converting it to a less harmful product?

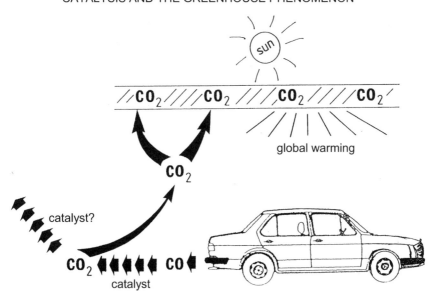

In the Discussion period, it was emphasised that CO_2 played an important role in the ICI process for methanol synthesis. Tom Wilkie, the Science Correspondent of *The Independent* newspaper, was present and the following day, 22 August 1990, the following headline appeared in the newspaper:

"Car exhaust gases could become fuel"

This is not exactly what the speaker had in mind! In 2007, we are faced with the potential severity of the problem encapsulated by the cartoon; what is as yet not clear is whether STM might provide the necessary molecular insight to designing an appropriate catalyst just as the Aarhus–Tøpsoe group achieved in their development of the nickel–gold steam-reforming catalyst.

The leading company involved in motor vehicle catalyst manufacture, Johnson Matthey, are clearly aware of the subtle chemistry revealed by nano-scale chemistry, with Golunski and Rajaram making the following observation (*Cattech*, 2002, **6**, No.1, 30):

> "... *Low-temperature catalysis can be unpredictable and transitory. It often arises from fragile interactions, which can be difficult to induce and even more difficult to control. In trying to understand these interactions, some of the long-accepted tenets of catalysis are brought into question.*"

It is, however, but one of the challenges that face catalysis in the future.

Subject Index

Acetylene, 151
Acetylene oxidation, 93
Adsorbate interactions, 65
Adsorption, 2, 14, 189
Alcohol, 92, 198
Alloys, 159, 193
Aluminium, 27, 53, 58, 73, 85
Ammonia, 24, 111
Ammonia oxidation, 24, 27, 80, 84, 138
Ammonia synthesis, 104
Antiphase domains, 184
Atomic force microscope (AFM), 33
Atom-tracking, 71
Auger electron spectroscopy (AES), 7, 19

Benzene, 93, 151, 160
Bimetallic, 173, 195
Buckled structures, 185

Carbon, 174
C60, 208
Carbon dioxide, 114
Carbon monoxide, 19, 27, 51, 64, 105
Carbon monoxide oxidation, 15, 51, 85, 118, 128, 163, 175
Carbon nanotubes, 213
Carbonate, 106, 113, 130
Catalytic oxidation, 77
Cesium, 2, 105, 107
Checker Board Model, 2, 4, 9
Chemisorptive replacement, 64, 183
Chiral surfaces, 208

Chlorobenzene, 148
Cicardi orni, 204
Cinchona alkaloids, 208
Clock structure, 187
Cluster compounds, 167
Cluster size, 157
Coadsorption, 23, 54, 56, 61, 77, 81, 85, 106, 130
Cobalt, 204
Complex formation, 25
Copper, 17, 24, 27, 53, 59, 79, 85, 92, 105, 110, 116, 124, 138, 185, 206
Corrals, 205
Cyclohexene, 128, 160

Dialkyldiacetylene, 211
Di-block polymers, 211
Dimethyl sulphide (DMS), 199
Disorder-order transition, 61
Dissociative chemisorption, 137, 145
Domain boundaries, 184, 208
Domain walls, 193
Dynamics of adsorption, 17, 21, 51

Electron spin resonance, 5
Electronic factor, 103
Eley-Rideal Mechanism, 8, 50
Enantioselective catalysts, 209
Ethane, 126
Ethene, 126, 150
Ethyne, 41
EXAFS, 18, 181

Field emission, 4, 6, 15
Fischer-Tropsch, 19, 103, 179
Free radicals, 5, 9
Frenkel equation, 21

Gold, 33, 113, 122, 131, 191, 195, 209
Gold clusters, 160, 164
Graphite, 160, 211

HAADF, 171
Heat of adsorption, 2, 22
Herringbone reconstruction, 190, 207
Hot atoms, 8, 21, 24, 54, 73
HREELS, 15, 23, 85, 117
HRSTEM, 171
Hydrodesulphurization, 169, 179
Hydrogen, 2, 7, 15, 123, 145
Hydrogen bonding, 84, 207
Hydrogen chloride, 55, 147
Hydrogen oxidation, 89
Hydrogen sulphide, 64, 182
Hydrogenation, 128
Hydrogen-deuterium exchange, 132

Imide coupling, 209
Inelastic tunnelling spectroscopy (IETS), 40
Infrared spectroscopy, 105
Insulator-metal transition, 176
Iron, 104, 163, 206
Iron oxide, 132, 165

Kinetics, 2, 13

Langmuir isotherm, 2
Langmuir model, 52
Langmuir-Hinshelhood, 8, 50
Lead, 23
Lennard-Jones model, 3, 13, 135
Lithography, 204
Lithium, 115
Low Energy Electron Diffraction (LEED), 1, 5, 7, 16, 107, 115, 143, 185, 192

Magnesium, 8, 24, 61, 83, 93
Magnesium oxide, 166
Mars-van Krevelen mechanism, 129
Melamine, 211
Mercaptan, 183
Mesoporous solids, 167
Methane, 159
Methanethiol, 195
Methanol, 19, 93
Methanol oxidation, 92
Methionine, 208
Micelle, 213
Mobility, 3, 4, 9, 53, 61, 64, 67, 82, 85, 128, 137, 190
Molecular beams, 51
Molybdenum sulphide, 171
Monte-Carlo simulation, 57, 79

Nanocasting, 213
Nanoclusters at oxides, 160
Nanoparticle geometry, 171
Nanowires, 211
NEXAFS, 207
Nickel, 3, 7, 15, 20, 27, 54, 56, 84, 90, 145, 150, 189
Nickel oxidation, 6, 16, 54
Nickel particles, 160
Nitric oxide, 19, 50, 124, 136, 139
Nitrogen, 6, 104, 142, 206
Nucleation, 6, 17, 163

Octanethiol, 197
Optical Simulation, 17
Oxidation, 51, 53, 57, 60
Oxide films, 32
Oxygen, 15, 20, 24, 51, 53, 60, 64, 66, 71, 108, 110
Oxygen activation: theory, 98
Oxygen transients, 8, 17, 20, 24, 57, 60, 99

Palladium, 61, 142, 147, 152, 167
Penning ionisation spectroscopy, 211
Perylene tetra carboxylic di-imide, 208

Subject Index 221

Phenyl iodide, 151
Photoelectron spectroscopy, 6, 8, 136, 150, 184
Photoemission, 6, 16, 18, 54
Phalocyanines: Copper, nickel, iron, zinc, 208
Physical adsorption, 2
Platinum, 7, 51, 70, 87, 126, 129, 190, 194
Poisoning, 179
Polyacetylenes, 212
Polyani-Wigner Equation, 14
Polymerization, 165, 211
Potassium, 104, 113
Propene, 72, 93, 104
Pyromellitic dianhydride, 209

Radicals, 27
Reflection absorption infrared spectroscopy (RAIRS), 8, 15, 210
Reforming, 180
Repulsive interactions, 4
Rhenium, 190, 194
Rhodium, 125, 193
Ruthenium, 17, 66, 87, 140, 190

S_8, 192
Scanning near field optical microscopy (SNOM), 36
Scanning tunnelling spectroscopy (STS), 38, 164
Self assembled monolayers, 203
SEXAFS, 18, 187
Silicon, 10, 32, 39, 149, 210
Silver, 27, 68, 86, 96, 104, 157, 189, 199, 208
Single atom positioning, 205
Step mobility, 55, 58, 131, 184, 198
Sticking probability, 5
Strontium, 166
Sulphur adsorption, 64, 179

Sulphur dioxide, 95
Sulphur trioxide, 97
Sum frequency generation (SFG), 15
Super-hydrophobic, 203
Surface buckling, 185, 189
Surface engineering, 203
Surface hopping (diffusion), 17, 21, 65, 138
Surface reconstruction, 6, 52, 118
Surface residence time, 3, 22, 65
Surface steps, 7, 132, 140
Surface stress, 184
Synchrotron Radiation, 19

Tartaric acid, 208
Temperature programmed desorption, 5, 14
Templating, 211
Tetraphalic acid (TPA), 207
Thiols, 195
Thiophene, 172
Titanium dioxide, 39, 162
Transients (see oxygen)
Transition state theory, 13
Tungsten, 2, 3, 6, 19
Two dimensional gas, 73

Ultra-violet photoelectron spectroscopy (UPS), 19

Vacancies, 146
Vibrational Spectroscopy, 14

Water, 20, 27, 50
Work function, 4, 6, 10, 15, 52

Xenon, 15, 205
X-ray photoelectron spectroscopy, 18

Zinc, 25, 56, 167